Saving Energy and Reducing CO$_2$ Emissions with Electricity

Saving Energy and Reducing CO$_2$ Emissions with Electricity

Clark W. Gellings, P.E.

River Publishers

Routledge
Taylor & Francis Group

LONDON AND NEW YORK

Published 2020 by River Publishers
River Publishers
Alsbjergvej 10, 9260 Gistrup, Denmark
www.riverpublishers.com

Distributed exclusively by Routledge
4 Park Square, Milton Park, Abingdon, Oxon OX14 4RN
605 Third Avenue, New York, NY 10017, USA

First issued in paperback 2023

Library of Congress Cataloging-in-Publication Data

Gellings, Clark W.
Saving energy and reducing CO$_2$ emissions with electricity / by Clark Gellings.
     p. cm.
Includes bibliographical references and index.
ISBN-10: 0-88173-667-8 (alk. paper)
ISBN-13: 978-8-7702-2295-2 (electronic)
ISBN-13: 978-1-4398-7012-9 (Taylor & Francis : alk. paper)
     1. Electric power--Conservation. 2. Industries--Energy conservation. 3. Carbon dioxide mitigation. I. Title.

TK4015.G36 2011
621.31'2132--dc22

2011000044

Saving energy and reducing CO$_2$ emissions with electricity / by Clark Gellings.
First published by Fairmont Press in 2011.

Routledge is an imprint of the Taylor & Francis Group, an informa business

Publisher's Note
The publisher has gone to great lengths to ensure the quality of this reprint but points out that some imperfections in the original copies may be apparent.

ISBN 13: 978-87-7022-910-4 (pbk)
ISBN 13: 978-1-4398-7012-9 (hbk)
13: 978-8-7702-2295-2   (online)
13: 978-1-0031-5164-7   (ebook master)

While every effort is made to provide dependable information, the publisher, authors, and editors cannot be held responsible for any errors or omissions.

Contents

Foreword

While this book was being readied for press, the U.S. Department of Energy's Energy Information Administration (EIA) released its short-term energy projections (EIA/STEO, October 2010). These projections are not good news for the U.S. economy and argue strongly for increased electrification.

It appears that 2010 will see the first increase in the U.S. energy intensity in many years. Historically, the rate of decline in the U.S. energy intensity has been around 1.5% per year over the last 50 years. In fact, in some years, the decline has exceeded 2.5%. EIA projections indicate an increase of 0.6% this year, driven in part by a 2.6% growth in gross domestic product (GDP). In part, the increase is also driven by decreasing investments and a reduction of asset utilization (e.g., factories reducing shifts).

This calls for an international imperative to make existing uses of electricity as efficient and practical as possible, as well as to electrify all inefficient fossil-fueled uses of energy.

Chapter 1

Introduction— Electricity's Attributes

Through different applications, electricity provides light, heat, comfort, and mechanical work. As long as society's desire for comfort, convenience and productivity is to be met, humans must find at least some reasonable quantity of energy forms which are accessible, affordable and have modest or zero environmental impacts. Without question, that will lead to electrification and a decrease in, and possible elimination of, the use of fossil fuels.

- It is only with electricity that we can fully leverage the use of renewable energy resources—resources which have zero carbon emissions.

- It is only by using electricity that we can effectively use nuclear power, biopower (electricity from sustainable biomass), hydro and kinetic (run-of-river, tidal and wave), and massive-scale geothermal energy resources which have zero carbon emissions.

- Only electricity has the advantage of providing access to the entire electromagnetic spectrum providing opportunities to use infrared, X-rays, ultraviolet rays, radio frequencies, microwaves, ultrasound, high-frequency AC for artificial illumination, Lorenz forces to turn motors, and other electrical phenomena which leverage energy input by up to several magnitudes.

- Only electricity can be precisely controlled and directed with near unlimited quantity with precision.

- Only electricity enables our digital economy and the communications and entertainment we have come to enjoy and depend on.

- Electricity is a refined energy form which can meet the world's energy needs in a sustainable manner. No other energy form has the range of attributes which electricity has.

Electricity is a uniquely valuable form of energy, offering unmatched precision and control in application as well as versatility and efficiency. Electricity also has the potential to provide superior environmental benefits when compared with other energy options. And electricity provides a clean, comfortable supply of energy. Because of these unique attributes, new electric appliances and devices typically require less total resources than comparable natural gas or oil-fired systems. Electric devices also provide greater performance and quality from an energy service perspective.

Electricity's utility is diverse. Certain energy forms can meet one need more efficiently than electricity, but these forms are extremely limited in their range of application. Only one energy form—electricity—can meet all of a customer's energy needs (comfort, convenience, appearance and productivity) as well as facilitate the achievement of other needs (medical diagnostics, money from automatic teller machines, personal computers, etc.). Electricity is unique in its ability to deliver packages of concentrated, precisely controlled energy and information efficiently to any point.

In addition, electricity can help alleviate many of the concerns facing the world (e.g., global warming, energy security, the use of limited resources, and the spiraling costs for obtaining them). In fact, electricity is uniquely suited for this critical task:

- It is available from various sources including low- or zero-carbon-emitting sources (nuclear, hydro, wind, solar, coal with carbon capture and sequestration) at a competitive cost.

- Its versatility allows it to be readily converted into easily and efficiently usable forms.

- Its efficiency at the point of end use is superior. Electricity's efficiency at the point of end use is substantially higher than fossil fuels due to the ability to leverage various portions of the electromagnetic spectrum.

Electricity offers society more than just improved energy efficiency. It also has greater "form value" than any other energy source: form value affords technical innovation with enormous potential for economic efficiency. Form value encompasses three dimensions: technical, economic, and resource uses.

During the 1880s, Thomas Alva Edison, the famous inventor, with a flamboyant reputation, was the industry's primary proponent of electrification. Edison promoted the idea that this technology provided opportunities for developing products that customers wanted and needed. Skeptics questioned whether Edison's research was valid. When one decried electric lamps as a "fraud upon the public," Edison responded by rigging a 3,000-ton steamship, the Columbia, with 115 Edison lamps. He then sent the ship on a journey around the tip of South America to San Francisco. Two and one-half months later, the ship arrived in San Francisco with half the lamps still working. The skeptics retreated, the press trumpeted Edison's achievement, and the viability of this new technology was established (Munson, 1985).

Early electrification was first applied to the electric telegraph, electro-plating, and arc lamp lighting systems. With limited yet important uses, electricity extended its possibilities through the discoveries made by scientists and technicians, as well as by inventor-entrepreneurs.

For industrial consumers in the late 1800s, electricity had a major competitive advantage in relation to steam: its transmissibility. In this form, energy could be transmitted and could supply an area that was separate from generation. By combining innovations of products and of technical processes, electricity was also an effective factor of organizational and commercial innovations (CIGRÉ, 2010).

Nikola Tesla, the discoverer of alternating-current (AC) transmission and many other electrical concepts and apparatus, was a visionary in looking at future applications of electricity. Among the concepts he envisioned in 1916 were (Tesla, 1916):

- Ship propulsion
- Agricultural applications for safeguarding forests against fires; the destruction of microbes, insects, and rodents
- Ship navigation including dispersing fog with electric force and illuminating oceans
- Load management
- Refrigeration
- Fountains enabled by electrical pumping
- Electric cooking
- Electric signs
- Instruments for building convenience
- Wireless transmission of information
- Telegraphic presentation of pictures

- Electrical typewriters prompted by voice command
- Electrotherapy
- Bathing with electricity
- Crime prevention
- Electric guns

ELECTRICITY POWERS GROWTH

A concept first introduced by Sam Schurr (Schurr, 1986) offers in-sight into the enormous value of electricity in our society. Schurr estab-lished a linkage between the growth of Gross Domestic Product (GDP) and the consumption of electricity. The hypothesis Schurr declared was that the U.S. economy is highly dependent on highly reliable and af-fordable digital-grade electricity. Electricity powers homes and busi-nesses and facilitates productivity. Electricity intensity—measured by the amount of electricity consumed per dollar of real GDP—has become a gauge of the nation's economic health.

Electricity demand remains sensitive to changes in economic growth. Growth in electricity use was in relative "lock step" with growth in the GDP from the end of World War II into the early 2000s. The tie be-tween electricity use and the economy was the product of many factors, including the development of electric technologies, economic activity, and the relatively stable price of electricity.

ELECTRICITY POWERS DIGITAL DEVICES

Electricity is the basis for telecommunication, the Internet, and the operation of all digital devices. One of the earliest commercial applica-tions of electricity was, in fact, the electric telegraph first demonstrated in 1837.

The importance of electrical power continues to grow as society becomes ever more reliant on digital circuitry for everything from e-commerce to industrial process controllers to the onboard circuitry in toasters and televisions. With the shift to a digital society, business ac-tivities have become increasingly sensitive to disturbances in the power supply. Such disturbances not only include *power outages* (the complete absence of voltage, whether for a fraction of a second or several hours),

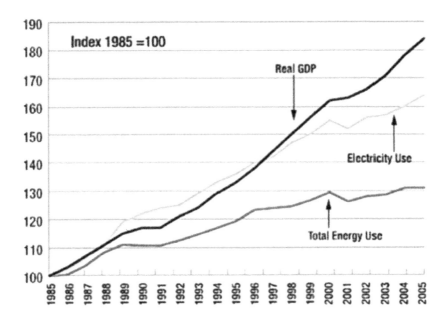

1985 represents the base year. Graph depicts increases or decreases from the base year.

Figure 1-1. U.S. Economic Growth is Linked to Electricity Growth (Source: U.S. Department of Energy, Energy Information Administration [EIA])

but also *power quality phenomena* (all other deviations from perfect power, including voltage sags, surges, transients, and harmonics).

Three sectors of the U.S. economy are particularly sensitive to power disturbances:

- **The digital economy (DE).** This sector includes firms that rely heavily on data storage and retrieval, data processing, or research and development operations. Specific industries include telecommunications, data storage and retrieval services (including collocation facilities or Internet hotels), biotechnology, electronics manufacturing, and the financial industry.

- **Continuous process manufacturing (CPM).** This sector includes manufacturing facilities that continuously feed raw materials, often at high temperatures, through an industrial process. Specific industries include paper; chemicals; petroleum; rubber and plastic; stone, clay, and glass; and primary metals.

- **Fabrication and essential services (F&ES).** This sector includes all other manufacturing industries, plus utilities and transportation facilities such as railroads and mass transit, water and wastewater treatment, and gas utilities and pipelines.

These three sectors account for roughly two million business establishments in the U.S. Although this is only 15% of all U.S. business establishments, these same three sectors account for approximately 40% of U.S. gross domestic product. Moreover, disruptions in each of these sectors—but especially DE and F&ES—have an almost immediate effect on other sectors that depend on the services they provide.

ELECTRICITY: GATEWAY TO
THE ELECTROMAGNETIC SPECTRUM

Electricity, or more precisely electromagnetic energy, is the only energy form which can provide a "gateway" to the electromagnetic spectrum. The electromagnetic spectrum is the range of all frequencies of electromagnetic radiation.

Electromagnetic waves have three physical properties: the frequency of the alternating cycle of the wave itself; the wavelength and the quantity of photon energy. As the frequency increases, the wavelength decreases. Photon energy increases proportionately to frequency.

The electromagnetic spectrum extends from below those used for electric power distribution (50 or 60 hertz (Hz)) to those used for AM/FM radio, television, microwave ovens, radar, infrared, visible light, ultraviolet light, X-rays and gamma rays.

Direct Current (DC)—While not actually part of the electromagnetic spectrum, direct current is a form of electricity which can stimulate electromagnetic waves. It is generated by chemical means in batteries by photovoltaic cells, fuel cells, and other generators. DC can also be derived from AC by use of rectifiers. DC is widely used in digital devices such as personal computers, desktop computers, fax machines, and portable electronics like mobile phones, personal data assistants (PDAs), and many other applications. DC is also used for high-voltage bulk power transmission. Many appliances and devices use DC somewhere internally in their circuitry.

Figure 1-2 illustrates the technical aspects of the electromagnetic spectrum.

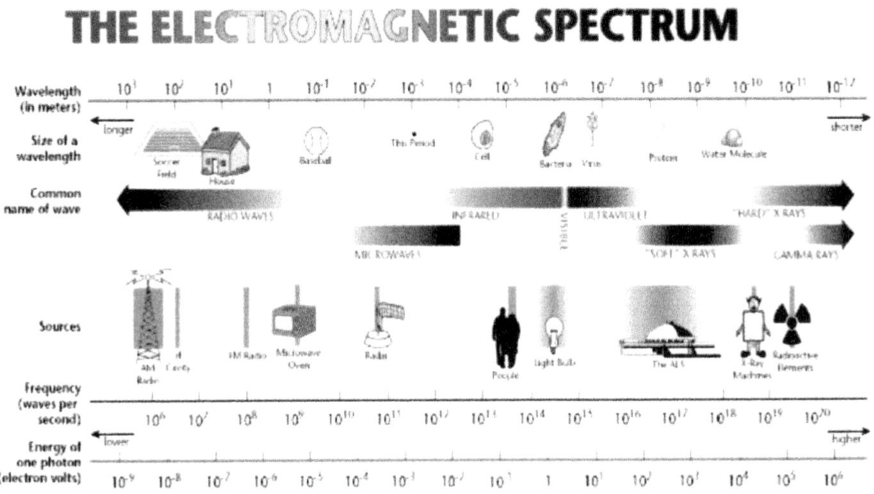

THE ELECTROMAGNETIC SPECTRUM

Figure 1-2. (Source:http://www.lbl.gov/MicroWorlds/ALSTool/EMSpec/EM-Spec.html.960105)

The electromagnetic spectrum consists of all forms of electromagnetic radiation, each corresponding to a different section, or band, of the spectrum. For example, one band includes radiation that our eardrums use to interpret sound. While another band, visible light, consists of radiation our eyes use to "see" light (see Figure 1-2). Notice how small the visible band is compared to the rest of the spectrum. An explanation of our eye's sensitivity to this small band described in the paragraph on Blackbody Radiation.

Power Distribution

Alternating current (AC) power distribution is generally at frequencies of 50 or 60 Hz. It is the frequency used to generate, transmit, distribute, and utilize power in modern power systems.* These frequencies also embody the utilization of the largest overall use of electricity—the electric motor.

Radio Frequency

Radio frequencies (RF) are those where radio waves are generated and propagated through the use of antennas. The size of antennas vary

*The reader should note that some limited power generation is DC and some high-voltage transmission is DC.

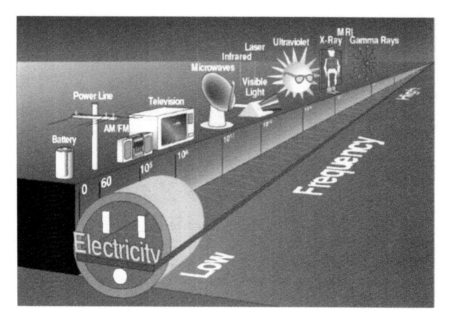

Figure 1-3. Gateway to the Electromagnetic Spectrum (Illustrated) (Source: C.W. Gellings, Electric Power Research Institute)

according to the frequency of the radio wave. RF waves can use materials as receivers or receptors. For example, a textile fabric in a dye bath can be subjected to RF to enhance dye penetration. More commonly, RF is used for transmission of musical voice or radios, data, TV, mobile phone calls, wireless networking, and amateur radio.

Microwaves

Microwaves (MW) are next up the scale on the electromagnetic spectrum. Microwaves employ special antennas which, because the waves are "short" enough, use tubes of metal as wave guides. Microwave energy is produced by Klystron or magnetron tubes or with solid-state devices. Microwaves are absorbed by liquids, like water that have dipoles. For example, food is heated in a kitchen using a microwave oven. The food is heated by exciting these dipoles. Low-intensity MW radiation is used in Wi-Fi as well. Volumetric heating can be done with microwaves. This literally transfers heat electromagnetically. As a result, MW heating can result in more uniform heating and reduces overall energy uses.

Terahertz Radiation

Terahertz radiation is a region of the spectrum between infrared and microwaves. Recently, applications such as imaging and communications are appearing in this frequency range. Scientists are also looking to apply terahertz technology in the armed forces, where high-frequency waves might be directed at enemy troops to incapacitate their electronic equipment (Wikipedia—Electromagnetic Spectrum).

Infrared Radiation (IR)

IR is typically categorized as far-infrared, mid-infrared, and near-infrared. Infrared radiation is an important means for heating humans and products in industrial and commercial settings. IR energy can be easily absorbed by many different colors and materials—much like IR from the sun. Therefore, it is universally applied in processes. It is the one component of the EM spectrum that can be generated by fossil fuels. Typically, gas is used to heat a material like a ceramic block or steel tube which then radiates IR to the surroundings. Reflectors can be used to help direct or focus the IR toward objects or work areas. Electric IR heating is a very popular method of industrial process heating.

Visible Radiation

Visible radiation is what we commonly refer to as light energy. Light is that portion of the EM spectrum where the sun emits most of its radiation. As such, it is natural that the human eye is most sensitive to these frequencies. This is another example of electricity's superiority as an energy form. A fossil-fueled flame cannot compare to the overall efficiency of an electric illumination system. Artificial illumination as lighting powered by electricity is used broadly in modern society.

Ultraviolet Radiation

Ultraviolet radiation (light) is just at the end of the visible spectrum. Best known as the spectrum most known to cause sunburn or even skin cancer, UV energy is very energetic and can break chemical bonds. It is often used to cure paints and special coatings.

X Rays

X rays are also very energetic—and have even higher energy levels which can cause them to alter materials. X rays can pass through materials and, as a result, are used in medical diagnostics (radiography). X rays are also used in high-energy physics for experimentation.

Gamma Rays

Gamma rays are the most energetic of all. They can be used to irradiate and sterilize food. They are also used to treat cancer patients, as they can be highly directed at tissues, such as a tumor.

TECHNICAL ATTRIBUTES OF ELECTRICITY

Electric-powered equipment offers significant advantages when compared with conventional (e.g., thermal) counterparts such as natural gas or oil including reduced energy use, increased productivity, and improved product quality, compactness, and environmental cleanliness. Electricity's technical attributes include electricity phenomena, input energy density, volumetric energy deposition, controllability, and synergistic combinations.

Residential appliances, building energy systems, and industrial processes frequently involve the interaction of energy and matter to modify materials, pump refrigerants or fluids, or to transform them from one form to another. Three types of electrical phenomena can be involved in these transformations: electromotive, electrothermal, and electrolytic. These phenomena are all unique to electricity as an energy form and contribute to its form value.

- Electromotive phenomena occur when mechanical motion is produced using electricity. The electric motor represents the most prominent example of electromotive phenomena, accounting for over 50% of the electricity consumed in industry. The electric motor is by far the most efficient and effective source of motive power: users can obtain efficiencies of over 90% with this device. The electric motor is used in a wide variety of applications including as "prime movers" (pumps, fans, and compressors), materials processors, and material handlers. In materials processing, for example, electricity can exert force without physical contact, permitting precise manufacturing of metal parts by rapidly accelerating them against a form.

- Electrothermal phenomena employ electricity to produce heat which, in turn, facilitates a physical or chemical change. Bulk processing industries (e.g., chemical, primary metal, stone, clay, glass, paper, and petroleum) transform material from one physical or chemical state to another. Three techniques exist: In the simplest method, direct ohmic dissipation, an electric current is passed through the material by physically attaching electrodes. The second method, electromagnetic induction, heats conducting materials without direct physical contact. The material to be heated is placed in proximity to a coil carrying an alternative current; the fluctuating magnetic field produced by the coil induces eddy currents in the material which are dissipated to produce heat. Finally, certain non-conducting materials can be heated dielectrically. This process occurs when a material containing polar molecules (e.g., water) is placed in a rapidly alternating electrical field. Dielectric losses produce heat as, for example, in a microwave oven.

- Electrolytic processes bring about chemical change through the direct use of electricity. Electrolytic phenomena occur at the molecular and atomic levels. The earliest (and still most common) industrial applications of electrolytic processes occur in the chemical industry. Electrolysis is best know for its use in the production of such basic materials as aluminum and chlorine.

Input Energy Density

In combustion processes using chemical fuels (e.g., oil and gas), the maximum achievable temperature is thermodynamically limited to the "adiabatic flame temperature," a practical limit of about 3000°F for fossil fuels burned in air. When heating material electrically, there is no inherent thermodynamic limit on the temperature. Typical temperatures of 10,000°F and higher are achieved routinely in arc-produced plasmas, and much higher temperatures are technically feasible.

Volumetric Energy Deposition

Electrothermal phenomena are volumetric (i.e., generating heat within the material itself). When using fossil fuels to heat material, heat is usually imposed at the surface by radiation and convection. This method is inherently slow and inefficient. With induction heating, electrical energy is deposited directly within the material; thus, processing time can be reduced to several minutes or less.

Volumetric heating can also be used to dry moist materials with microwave or radio frequency radiation. In conventional drying, heat must diffuse into the material from the surface, while moisture diffuses out, a slow process since most materials of interest are poor thermal conductors. If drying is accelerated too much by intensifying the rate of heating, overdrying of the surface can occur, causing cracking and degradation of the product. Dielectric heating largely eliminates this problem by greatly increasing drying rates. Thus, improving overall productivity, product yield, and product quality.

Controllability

Electricity is an "orderly" form of energy, in contrast to thermal energy which is random in nature. This means that electrical processes can be controlled much more precisely than thermal processes. Since electricity has no inertia, an industry can instantly vary energy input can be varied in response to process conditions, such as material temperature, moisture content, or chemical composition while accurately maintaining a desired state. Lasers and electron beams can be focused on a work surface to produce energy densities a million times more intense than an oxyacetylene torch. The focal points of these high-intensity energy sources can be rapidly scanned with computer-controlled mirrors or magnetic fields to deposit energy exactly where it is needed. This focusing capability offers a tremendous advantage, for example, in heat-treating a part precisely at points of maximum wear, thereby eliminating the need to heat and cool the entire object. Electrolytic processes impart energy directly to ionic species to produce molecular separation or selectively induce chemical reactions.

Synergistic Combinations

In some processes, electrolytic, electrothermal, and electromotive effects combine in an advantageous way. In the Hall-Heroult process for reducing alumina to aluminum, ohmic heating helps keep the cryolite bath in the molten state, while electrolysis causes pure aluminum to separate out and collect at the cathode. In a coreless induction melting furnace, electromagnetic induction heats and melts the charge while at the same time inducing a strong electromotive stirring action, which enhances heat transfer to the solid material and greatly improves homogeneity of the melt. The latter effect is especially important in the production of high-alloy castings and can be a major determining factor in the choice of induction melting over alternative methods.

ECONOMIC ATTRIBUTES

Economic attributes include fixed cost, flexibility of raw material base, product quality and yield. Some economic attributers are considered by economists as "externalities." These can include societal impacts like environmental emissions, energy security, and macroeconomic factors.

High-energy density and precise control present in electric systems typically result in increased production rates which, in most cases, reduce fixed costs per unit of product. With, for example, faster throughput, components such as labor, overhead, and interest on capital are spread over a larger production volume. Thus, even when the cost premium of electricity increases the energy cost per unit, total production cost per unit often remain lower.

Electrical process equipment is typically more economical in smaller unit sizes than combustion equipment, since it does not require fuel handling and environmental control. Electrical production technology can give rise to an "economy of reduced scale," in that smaller equipment requires less space, and industries can decentralize facilities and site them in proximity to diverse raw material sources and product markets. This flexibility, in turn, carries with it great economic benefit.

Flexibility of Raw Material Base
The high-energy intensity and precise control offered by electro-technologies permit a greater degree of flexibility with regard to raw material resources than do fossil-fueled processes. Arc furnaces, for example, can utilize either scrap or direct-reduced iron as a basic resource with little or no process modification.

Product Quality and Yield
In overall production economics, product quality and yield are critically important. Electrical processes typically provide improvements in both areas.

Resource Use
Resource use attributes of electric processes include use of renewable resources; flexibility of fuel supply; domestic resource balance of payments; national security; environmental; and energy consumption.

Use of Renewable Resources

Electricity offers the opportunity to use a number of low or zero carbon-emission sources of electric power production. Nuclear power production, coal- or gas-fired generation with carbon capture and sequestration, wind, solar, geothermal, biopower and hydro and kinetic power production are expanding in overall availability to reduce and nearly eliminate CO_2 and other emissions.

Flexibility of Fuel Supply

Combustion-based processes are highly dependent on the availability of specific fuel sources, since combustion equipment in general is not very adaptable to changes in fuel type. The shift to electricity-based processes delegates responsibility for fuel choices to the electric wholesale market utility, which can optimize fuel diversity. In addition, efficient energy conversion is a power generator's primary concern. Thus, over the long term, basic manufacturing processes or business activities can remain essentially the same while the electricity markets respond to changes in primary fuel markets.

Domestic Resource Balance of
Payments and National Security

Total resource requirements, and therefore, the need for imported fossil fuels, declines with the increased use of electricity due to the overall efficiency of electricity-based systems and the use of low and zero carbon power generation. As these imports decline, the impact on a nation's balance of payments may decline as do the risks in protecting national security.

Environmental

Electric processes and systems are undisputedly the most environmentally benign at the point of end use. Recently, it has also become evident that the application of electrotechnologies can mitigate what would otherwise be adverse environmental impacts in transportation, industrial, and in municipal and medical waste applications. Even when converting raw resources to produce it, electricity becomes the environmentally preferred energy form due to its efficiency at the point of end use. With increasing use of renewable power generation and other low-carbon-emitting generation, electricity is the most environmentally friendly energy form.

Energy Consumption

Primary energy consumption is almost always lower for electrical processes than with conventionally fueled systems. Less than 70% of electricity generation is from coal and gas generation. And the proportion of this use will decrease over time. The balance of resources (hydro, nuclear, and renewables) have virtually 100% efficiency from a carbon perspective.

Electricity Leverages Exergy

According to the second law of thermodynamics, the quality of energy is degraded every time it is used in any process. The term used to define the measure of energy quality has come to be knows as exergy. The term was coined in 1953 by Zoran Rant (www.eoearth.org). The theory is based on flows of energy between a subsystem or process and the systems surrounding that subsystem. Once energy exchanges between these systems are in balance—no more useful work can be done. Exergy, therefore, is the maximum useful work that can be extracted from a system until it reaches equilibrium with its environment.

The exergy associated with a transfer of a stock of energy can be defined as the potential of maximum work which could ideally be obtained from each molecule of energy being transferred or stored.

The importance of this concept cannot be overstated. The first law of thermodynamics teaches that energy cannot be created or destroyed. However, all of the solar system's resources will lose energy each time they are converted. Even through natural means, the high energy from the sun is converted to grow plants and feed humans or animals. The exergy lost in that process cannot be replaced even if the energy developed in humans is converted to useful work and human waste is recycled.

The first law of thermodynamics tells us that the amount of energy in the universe remains constant. But, for example, once coal is extracted from the earth and consumed in the production of electricity, the heat and emissions rejected in its production and any work actually done once the electricity is delivered and again converted to useful work can never be reversed. The exergy in coal has been converted to electricity and then, once again, into work (e.g., motive power).

If we could convert coal, for example, directly into useful work—instead of the sequence coal:fire:steam:electricity:motor:mechanical work—we could go directly from coal to mechanical work in a process and we would maximize exergy.

Since we have not perfected a process to convert coal directly to electricity, a more practical example would be that of combined heat and power (CHP) system, also called cogeneration. CHP systems leverage the otherwise rejected heat from fossil-powered generation and use that heat to power processes, heat or cool buildings, or provide domestic hot water. In this way, they capture more exergy than in a conventional process.

Figure 1-4 illustrates the mass-flow of a typical process. An industrial process was used for this example. In this diagram, exergy is optimized anytime the number of conversions of energy are minimized.

Electricity, as an energy form, is the most optimal way to extract the most exergy possible in the utilization of basic energy. For example, in using heat pumps to either heat or cool a building or a process, electricity allows greater exergy by leveraging renewable sources of generation and heat pump technology which extracts heat from its surroundings or from waste heat. It has been suggested that a more appropriate method to measure energy efficiency is to measure its exergy (Favrat, 2010).

Figure 1-5 compares the exergy efficiency of two alternative building domestic water heating systems. Each assumes the use of natural gas which is extracted from the ground, compressed (using natural gas)

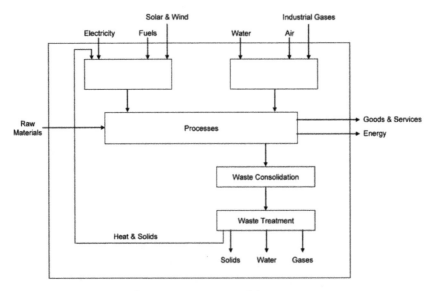

Figure 1-4. Energy and Processes

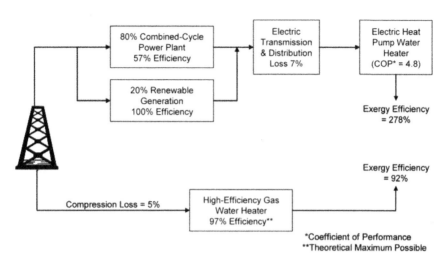

Figure 1-5. Alternatives for Electric vs. Gas Water Heating

and delivered either to a combined-cycle power plant or directly to a high-efficiency gas water heater.

In the electric example, it is assumed that the fossil-powered plant delivers only 57% of the electricity need. The balance is provided by 100% efficient renewable resources such as wind, solar or hydro. In this example, after electricity T&D losses are applied, the electricity is used to energize a heat pump water heater with a theoretical coefficient of performance (COP) of 4.8.

Both the electric and gas water heater efficiencies higher than currently available on the market. However, laboratory tests have verified that electric heat pump water heaters using CO_2 as a refrigerant could achieve COPs of as high as 5.2. In the gas of the gas water heater, the efficiency assumption is based purely on theoretically achieved potential.

This example illustrates the extraordinary potential which advanced electric technologies can have in leveraging exergy.

A similar example can be offered in technologies such as freeze concentration and the use of heat recovery chillers.

Examples of electric end-use technologies which maximize exergy include:

- Heat pumps (air to air, water loop, geothermal, heat recovery, etc.)
- Heat recovery chillers
- Variable-speed industrial compressors

- Hermetic turbocompressors or turbines—direct drive, two-stage
- Fourth-generation nuclear reactors
- Fuel cells
- Freeze concentration
- Microwave heating
- Ultraviolet curing
- Infrared Heating
- Radio-Frequency Drying

References
"Electricity and Industrial Productivity," EPRI, Palo Alto, CA: 1984. EM-3640.
"The Power Makers," R. Munson, 1985, p. 46.
"The History of CIGRÉ," CIGRÉ, Paris, France, 2010, pending publication.
"Electricity Use, Productive Efficiency and Economic Growth," S. Schurr and S. Sonenblum, EPRI, Palo Alto, CA: 1986.
"Industrial Energy Futures: Some hints on indicators, methods and technologies," D. Favrat, Ecole Polytechnique Féderale de Lausanne, presentation at U.S. DOE Energy Futures Study Workshop, July 28, 2010.
"Wonders of the Future," N. Tesla, *Collier's Weekly*, December 2, 1916.

Chapter 2

The Concept of Electrification

Numerous studies have highlighted the fact that society's thirst for energy can best be met with electricity. This stems from two drivers: the increasing effectiveness of new electricity generation as a low-carbon energy source, and the fact that electricity at the point of end use can have a leveraging effect which results in delivering many times more usefulness than fossil-fueled devices. The increasing effectiveness of new generation stems from the fact that electric generation is increasingly from low-carbon-emitting technologies like coal with carbon capture and sequestration, nuclear, wind, solar, biopower, geothermal, water power, and other effective means.

Figure 2-1 illustrates the importance which the efficiency at the point of end use has on carbon emissions for a given application. The figure illustrates three technologies which can be used for domestic water heating: electric resistive water heaters, gas-fired water heaters, and heat pump water heaters. In this illustration, the annual carbon emissions of typical residential water heaters are compared to both the energy factor (EF) and the carbon intensity of delivered electricity in pounds per kilowatt hours. The electric heat pump is a superior technology across the range of carbon intensity for electricity. However, most notable is the fact that as the carbon intensity of the electric supply mix decreases, the overall effectiveness of the advanced electric heat pump technology increases dramatically.

Figure 2-2 illustrates the CO_2 emissions intensity for a low-carbon supplier. As the portfolio of generation options continues to have a lower carbon footprint, such as in the case of Exelon, electrification will become the preferred strategy to reduce overall carbon emissions.

Figure 2-3 illustrates the opportunity which electric end-use technologies can offer in reducing CO_2 as compared to gas end-use technologies. In this illustration, as overall CO_2 emissions per kWh decrease,

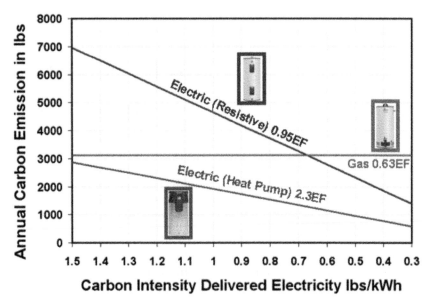

Figure 2-1. Carbon Emissions Water Heater Technology

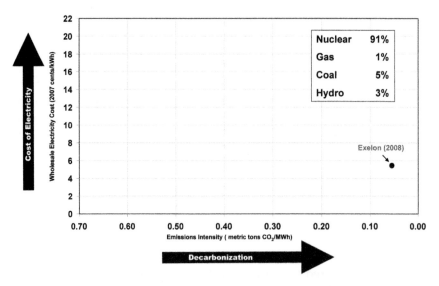

Figure 2-2. CO_2 Emissions Intensity vs. Wholesale Electricity Cost for a Low-Carbon Supplier

electricity is the preferred option. Likewise, as the coefficient of performance (COP) of electric technologies increases, so does the ability of electricity to displace carbon. Thus, the combination of de-carbonizing electricity coupled with the increasing effectiveness of electric end-use technologies points toward electrification as the preferred option.

Figure 2-4 illustrates CO_2 emissions from a range of electric and gas technologies. Overlaid on the graph is today's average carbon intensity of the electricity sector and the CO_2 reduction targets often mentioned in political debates (3% by 2012, 17% by 2020, 42% by 2030, and 83% by 2050). In just six years or less, electric heat pump water heating will become the most effective technology to heat water from an energy and carbon perspective.

Note that a COP = 4.8 may be a little too optimistic for heat pump water heaters (HPWHs) applied in the average U.S. household—a COP = 4.2 may be more realistic in the short term.

Figure 2-5 illustrates the actual break-even points between the de-carbonization of power production (measured in decreasing CO_2/kWh) and the increasing effectiveness of electricity end use (measured in increasing COP). This illustration compares HPWHs with the most efficient Energy Star gas systems available (condensing, tankless, 0.98EF). It plots a lower-bound COP of 2.0 to represent a standard HPWH. Beginning with a smaller COP value on the axis would make the triangle larger.

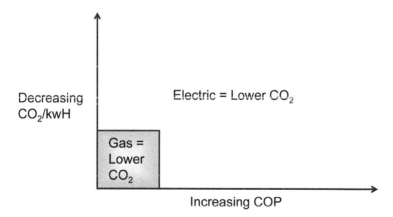

Figure 2-3. Decreasing CO_2/kWh Emissions From Electric Generation Coupled With the Increasing Effectiveness of Electricity End-Uses Leave Few Areas Where CO_2 Emissions From Gas Are Lower

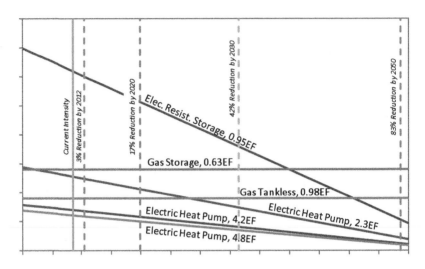

Figure 2-4. Carbon Footprint for Various Water Heater Technologies

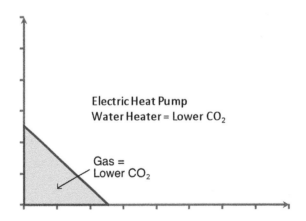

Figure 2-5. Heat Pump Water Heaters Become Increasing the Preferred Choice

Figure 2-6 illustrates a similar analysis for heat pump space conditioning. Once again, well before 2020, electric heat pumps will outpace gas heating as the most effective means to meet currently debated CO_2 reduction targets from the electricity sector.

Figure 2-7 summarizes the heat pump's effectiveness in space heating and cooling as electricity's carbon content decreases and electricity end uses become more effective. This compares heat pumps with the

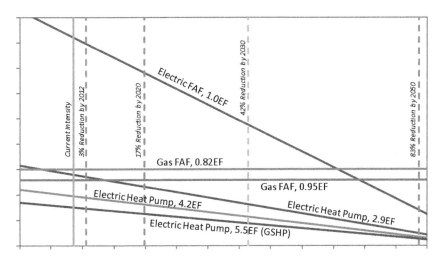

Figure 2-6. Carbon Footprint: Space Heating Technologies

most efficient Energy Star gas forced air furnace (FAF) systems available (condensing, 0.95EF). It shows a lower-bound COP of 2.5 to represent a standard heat pump for space heating. Since heat pumps can be more efficient for space heating than water heating under average U.S. conditions, the gas triangle is smaller for the space heating slide. Again, changing the minimum COP value would increase the triangle size.

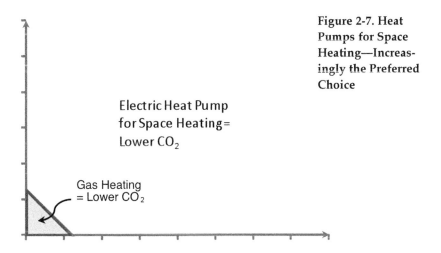

Figure 2-7. Heat Pumps for Space Heating—Increasingly the Preferred Choice

EPRI'S PRISM AND MERGE

In 2007, the Electric Power Research Institute (EPRI) released its first Prism and MERGE analyses, providing a technically and economically feasible roadmap for the electricity sector as it seeks to help reduce greenhouse gas emissions over the next few decades. The Prism analysis provided a comprehensive assessment of potential CO_2 reductions in key technology areas of the electricity sector. The MERGE analysis identified the economically optimum technology portfolio for a given CO_2 emissions constraint.

The 2009 update reflects economic and technological changes that have the potential to affect projected emissions and the technologies to address them. The 2009 update included two electrification strategies: one for electric transportation and the second for beneficial new uses.

The Prism analysis determined that the sector can potentially meet the challenges that confront each technology option and deploy "The Full Portfolio" of technologies to achieve substantial emissions reductions. The top line of the graph (Figure 2-8) represents the U.S. Energy Information Agency's (EIA) 2009 Annual Energy Outlook reference case estimate of CO_2 emissions from the U.S. electricity sector through 2030. Each color represents the incremental reduction in emissions projected as feasible for a given technology. The Prism analysis illustrates the overall reductions achievable using The Full Portfolio of technologies.

The MERGE analysis determined the most economic combination of these technologies over time to meet a specified CO_2 emissions constraint. MERGE projects electricity generation from different technologies, electricity costs, CO_2 prices, and the overall cost of implementing CO_2 emissions reductions.

The Prism analysis estimates that the technical potential exists for the U.S. electricity sector to reduce annual CO_2 emission in 2030 by:

- 41% relative to 2005 emissions based on improvements to electric sector technologies;

- 58% relative to 2005 emissions, if reductions due to electrotechnologies and electric transportation are included; and

- 62% relative to the 2030 reference case projection in the Energy Information Administration's 2009 Annual Energy Outlook (EIA 2009).

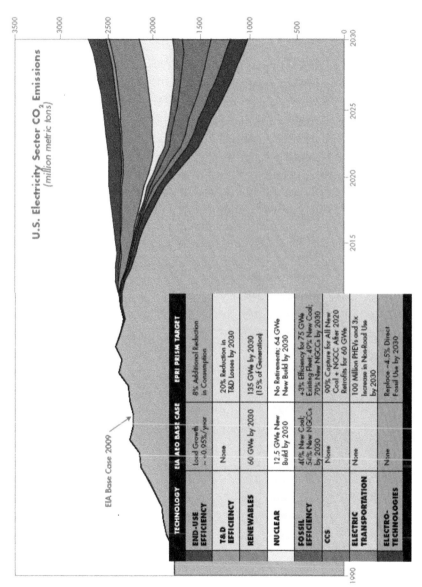

Figure 2-8. EPRI's Prism Analysis (Source: EPRI)

The Prism analysis projects that by 2030, 60% of the total U.S. generation mix would consist of low- or non-CO$_2$-emitting generation—provided that the required research, development, and technology demonstrations can be carried out and the technical assumptions can be met.

ELECTRICITY TECHNOLOGY UNDER A
CARBON-CONSTRAINED FUTURE

Based on a "bottoms-up" review of technology performance capabilities and deployment potential, EPRI has developed a technical assessment of the feasibility for future U.S. electricity sector CO$_2$ emissions reductions. This so-called "Prism" analysis (from the colorful appearance of the graphical results) represents an estimate of the potential electric sector CO$_2$ emissions reductions, in that it focuses solely on technical capabilities assuming no economic or policy constraints. It also demonstrates that electrification reduces CO$_2$ emissions.

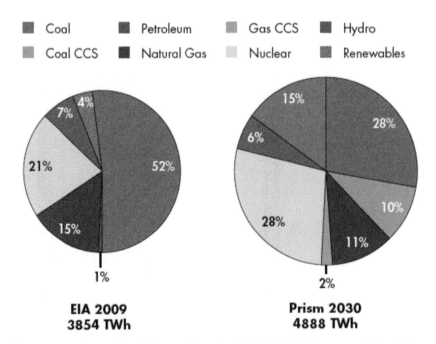

Figure 2-9. Generation Mix as Contained in EIA's Estimate and the Prism Analysis

The Technology Portfolio

The Prism analysis assumes successful achievement of performance and deployment targets associated with several advanced technologies as a basis for estimating CO_2 emissions reduction potential:

- End-use energy efficiency
- Renewable energy
- Advanced light water nuclear reactors
- Advanced coal power plants
- CO_2 capture and storage
- Plug-in hybrid vehicles
- Distributed energy resources
- Electrification

The technologies considered and the selection of "aggressive but feasible" analysis targets were based on capabilities that still face substantive research, development, demonstration, and/or deployment challenges, but for which a specific sequence of RD&D activities can be identified that will achieve wide-scale deployment of the technologies between today and 2030. Capabilities requiring assumption of breakthrough technology developments or which have deployment timelines past 2030 were excluded from the analysis.

The 2009 analysis estimates a potential CO_2 emissions reduction in 2030 of 9.3% as a result of electricity displacing gasoline and diesel to fuel a substantial portion of the vehicle fleet.

The 2009 Prism bases this estimate on the assumption that plug-in hybrid electric vehicles (PHEVs) are introduced to the market in 2010, consistent with product plans of many automakers, and the subsequent rapid growth of market share to almost half of new vehicle sales within 15 years.

The 2009 analysis estimates a potential CO_2 emissions reduction in 2030 of 6.5 as a result of electric technologies displacing traditional use of primary energy consumption for certain commercial and industrial applications.

Electrotechnology research (EPRI 1018871) indicates there are applications through which net reductions in CO_2 emissions can be achieved. This projection is based on replacing significant use of direct fossil-fueled primary energy with relatively de-carbonized electricity for selected applications: heat pumps, water heaters, ovens, induction melting, and arc furnaces.

2009 Prism Analysis Assumptions
- 4.5% of primary energy supplied by fossil fuels is replaced by electricity by 2030.

MERGE ANALYSIS

The 2009 MERGE analysis estimates the economically optimum portfolio of electricity sector technologies that will meet a CO_2 emissions constraint comparable to those suggested in current policy proposals. MERGE (Model for Estimating the Regional and Global Effects of Greenhouse Gas Reductions) analyzes the economy-wide impact of climate policy (EPRI 1019563).

The 2009 MERGE analysis compares the technology scenarios: "limited portfolio" and "full portfolio." The full portfolio assumes availability of CCS, advanced nuclear, significant improvement in costs of renewables, availability of plug-in hybrid vehicles (PHEVs), and accelerated improvements in end-use efficiency. The 2009 analysis includes the following:

- Emissions constraints indicative of current U.S. and international policy proposals (80% below 2005 levels for developed countries);

- Unconventional resources such as shale gas factored into natural gas supply;

- CCS retrofit for up to 60 GW of existing coal plants;

- Grid integration costs considered for high levels of variable output generation from renewables; and

- Higher biomass feedstock costs for large-scale biofuels and/or biomass electricity production.

Under CO_2 emissions constraints representative of current proposals, MERGE projects that the economically optimal full technology portfolio consists of substantial amounts of renewable electricity generation, significant electricity production from coal and nuclear, as well as large reductions in electricity consumption. Retrofit of CO_2 capture and storage for existing coal plants plays an important transitional role between 2010 and 2030. The sharp growth of new coal with CCS after 2030 will be driven by continually tightening emissions constraints, retirement of

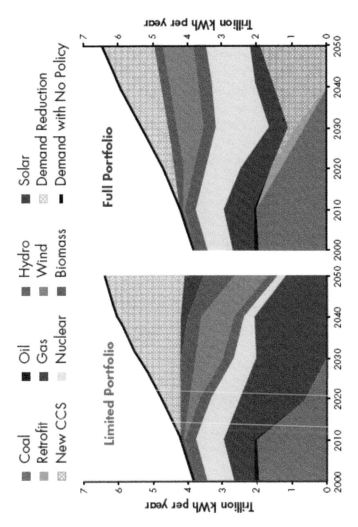

Figure 2-10. The Prism/MERGE Analysis of Future Impacts

coal units with CCS retrofits, the need to reduce emissions from natural gas, and anticipation of more limited uranium supplies for nuclear plants based on the once-through fuel cycle.

MERGE results indicate that the Prism technology assumptions are reasonable; comparison of the MERGE and Prism technology mixes in 2030 validates this. The generation shares of the different technologies in MERGE are generally consistent with those in the Prism. This suggests that the Prism technology portfolio may also be economically optimal.

The 2009 Prism analysis estimates that the technical potential exists for the U.S. electricity sector to reduce annual CO_2 emissions in 2030 by:

- 41% relative to 2005 emissions, based on improvements to electric sector technologies;

- 58% relative to 2005 emissions, if reductions due to electrotechnologies and electric transportation are included; and

- 62% relative to the 2030 reference case projection in the Energy Information Administration's 2009 Annual Energy Outlook.

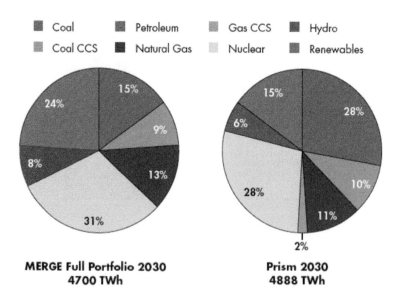

MERGE Full Portfolio 2030
4700 TWh

Prism 2030
4888 TWh

Figure 2-11. Prism and MERGE of 2030 Generation Needs

The 2009 Prism and MERGE analyses underscore the importance of research, development, and demonstration leading to a full portfolio of electricity sector technologies.

The full portfolio comprises both supply- and demand-side technologies: end-use efficiency and plug-in hybrid electric vehicles supported by a smart grid; wind, biomass, solar, advanced nuclear, and coal with CO_2 capture and storage.

The MERGE analysis indicates that a technology portfolio similar to that outlined by the Prism can achieve CO_2 emissions reductions at a considerably lower economic cost—as much as $1 trillion in some scenarios. Much of the required technology is not yet available, and substantial, sustained research, development and demonstration is required. Low-carbon electricity technologies drive growth in electricity demand even as CO_2 emissions are reduced.

This full portfolio analysis highlights the opportunity to further electrify the nation by leveraging an enhanced low-carbon generation portfolio.

The electricity sector will inevitably play a key role in new climate policies—both because power plants represent a major source of carbon dioxide and because electricity generation provides the most promising way to utilize a variety of primary energy resources including some that do not emit greenhouse gases (GHGs). In addition, advanced power generation technologies will sequester the CO_2 produced by burning fossil fuels, thus reducing atmospheric accumulation of this key GHG while preserving the option of continuing to use critical energy resources (e.g., coal).

EPRI's Climate Contingency Roadmap Phase I report (EPRI 1009311) first quantified the insight that more rapid electrification is crucial in a carbon-constrained world. Based upon the analyses in that report supplemented by historic data, "Electricity's role in the energy system is expected to continue growing as it has over the last 50 years and, indeed, the last century. This increased electrification has led to a gradual de-carbonization of the global energy system because the vast majority of non-emitting energy technologies are associated with electricity. In the U.S., for example, the carbon intensity of power generation has fallen by 10% over the last two decades due to the increased use of nuclear power and a shift from coal to natural gas."

The main reason behind this increasing electric intensity as a response to carbon constraints is that electricity is currently the only way to provide many critical energy services without emitting GHGs. Approximately 30% of U.S. electricity is generated by non-emitting sources, compared to just 3% of the energy used in transportation; and, in the future, non-emitting renewable and nuclear resources will likely play an increasing role in power generation. In addition, economics of scale and the fixed nature of generation facilities make deploying carbon reduction technologies at power plants cheaper than trying to apply such technologies to millions of small, dispersed emissions sources (e.g., vehicle engines and home furnaces). In order to reap the full benefits of electrification, however, an integrated approach for developing new technologies will be needed and climate policy must allow and encourage economically efficient shifting of energy use and carbon emissions among various sectors of the economy.

EUROPEAN CLIMATE FOUNDATION

Figure 2-12 illustrates the results of a study recently completed by the European Climate Foundation (European Climate Foundation, 2010). It highlights the potential to reduce GHG by 80% through the increased use of renewable energy coupled with "fuel shift" from conversion of fossil-fueled end uses to electric.

EURELECTRIC

A study by Eurelectric, the European Union of the Electricity Industry, found similar results. Entitled "The Role of Electricity," the study envisaged the use of all options towards a low-carbon energy system—energy efficiency, renewables, nuclear energy, and CCS. The scenario exploits the synergy between a low-carbon electricity supply system and efficient electrotechnologies including in areas traditionally largely limited to direct combustion of oil and gas—namely road transport (Eurelectric, 2007).

Eurelectric found that by electrification they could control both total energy costs and oil and gas import dependency while achieving carbon reduction without additional total energy costs. Oil and gas import dependency in 2030 and 2050 remains almost stable compared to 2005, whereas all other studies have identified a rise in oil and gas depen-

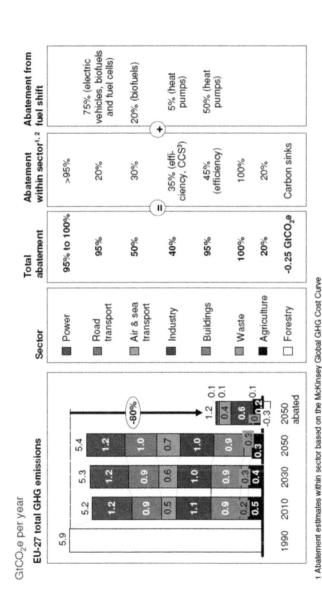

Figure 2-12. 80% De-Carbonization Overall Means Nearly Full De-Carbonization in Power, Road Transport, and Buildings (Source: McKinsey Global CHC Abatement Cost Curve: IEA WEO 20098; U.S. EPA; Team Analysis)

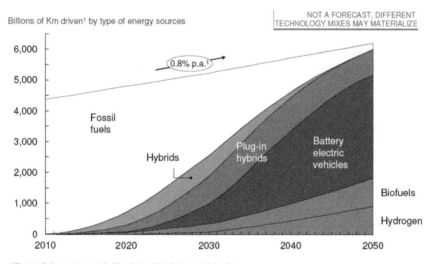

Figure 2-13. The De-Carbonized Pathways Assume a Mix of Electric Vehicles, Biofuels, and Fuel Cell Vehicles (Source: Team Analysis)

dency. In addition, electrification leads to a reasonable and stable level of carbon value—some €40-50/tCO_2—whereas other proposals peak at €120/tCO_2.

Eurelectric also found that electrification provided a robust energy future due to its inclusive portfolio. It is less vulnerable to unforeseen events and can better exploit opportunities offered by technological development. The portfolio approach is based on reasonable development of all options.

In particular, the Eurelectric study identified massive synergy between a low-carbon electricity supply and energy-efficient electrotechnologies, which constitutes an essential part of the solution. This synergy is important in all sectors of the economy but is particularly significant for two sectors that have until now been highly dependent on the direct combustion of oil and gas—namely space heating and transportation. For example, by 2030 Eurelectric found that heat pumps driven by the right electricity supply mix can heat a house or office at a small fraction (around 80% reduction) of the CO_2 emissions made by current oil or gas heating, simultaneously reducing oil/gas consumption by 90%. On the same time horizon, driving a plug-in hybrid car entails a spectacular (circa 70%) reduction in CO_2 emissions, while reducing oil/gas consumption by some 80%.

Table 2-1 summarizes the results of the Eurelectric study as compared to a baseline and two alternative scenarios—one relying on supply-side energy efficiency and renewable energy resources (RES).

The International Energy Agency (IEA) believes that a global energy technology revolution to meet climate change and energy security challenges is needed. While they point to some early signs of progress, they believe much more needs to be done. However, three questions remain (IEA, 2010):

- Which technologies can play a role?
- What are the costs and benefits?
- What policies are needed?

As shown in Figure 2-14, IEA believes that a basket of near-zero carbon-emitting technologies are needed to meet society's future needs. These include:

- Coal and natural gas with carbon capture and sequestration (CCS)
- Renewables
- Nuclear
- Power generation efficiency and fuel switching
- End-use fuel switching
- End-use fuel and electricity efficiency

Table 2-1. Summary of Energy System Changes (2050)

Scenario Results for 2030	Baseline	Electrification	Supply Scenario	Efficiency & RES
Final Energy Demand (2005=100)	118	106	113	102
Electricity Consumption (2005=100)	145	172	143	127
Electricity Price (2005=100)	111	121	133	123
Electricity from Nuclear (TWh)	654	1,643	1,535	852
Electricity from Renewables (TWh)	1,092	1,359	1,267	1,675
CO_2 Stored (cumulative Mt)	---	3,797	5,315	---
Power Investment (cumulative GW)	928	1,090	950	984

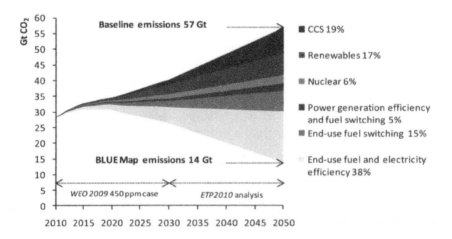

A wide range of technologies will be necessary to reduce energy-related CO_2 emissions substantially.

Figure 2-14. Key Technologies for Reducing Global CO_2 Emissions (Source: www.iea.org/techno/etp/index.asp)

IEA investigated a number of scenarios under what it referred to as a "new age of electrification." As shown in Figure 2-15, each scenario highlighted that a mix of renewables, nuclear, and fossil-fuels with CCS will be needed to de-carbonize society through de-carbonizing the electricity sector. The "Blue Map" scenarios contain aggressive electrification coupled with increased de-carbonization of the electricity sector.

CONCLUSIONS

Numerous studies highlighted in this chapter substantiate the fact that a low-carbon energy future can only be achieved if the following three dimensions, illustrated in Figure 2-16 are pursued.

• Maximize the end-use efficiency of electricity (and any fossil-fueled uses). This includes the end use of solar and geothermal systems for building and domestic water heating and cooling.

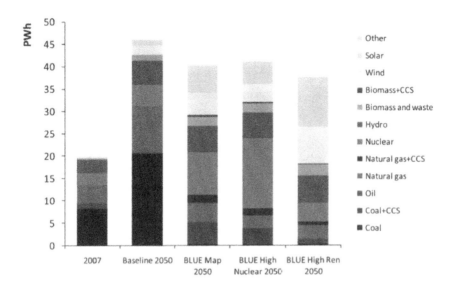

A mix of renewables, nuclear and fossil-fuels with CCS will be needed to decarbonise the electricity sector.

Figure 2-15. De-Carbonizing the Power Sector—A New Age of Electrification? (Source: www.ieg.org/techno/etp/index.asp)

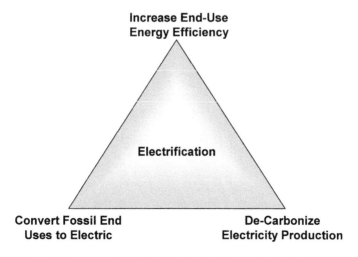

Figure 2-16. Electrification to Reduce CO_2 Emissions

- De-carbonize electricity production through the use of hydro, wind, solar, nuclear, geothermal, biopower, hydro, and kinetic energy and carbon capture and sequestration applied to coal-fired power plants (and to any gas-fired power plants which still exist).

- Electrify most all end uses of energy.

References

"Roadmap 2050: A Practical Guide to a Prosperous, Low-Carbon Europe," European Climate Foundation, April 2010.

"Climate Contingency Roadmap: The U.S. Electric Sector and Climate Change," EPRI, Palo Alto, CA: 2003. 1009311.

"An Updated Annual Energy Outlook 2009 Reference Case Reflection Provisions of the American Recovery and Reinvestment Act and Recent Changes in the Economic Outlook," U.S. DOE Energy Information Administration (EIA), April 2009, SR/OIAF/2009-03, www.eia.doe.gov.

"The Potential to Reduce CO$_2$ Emissions by Expanding the End-Use Applications of Electricity," EPRI, Palo Alto, CA: 2009. 1018871.

"Prism/MERGE Analysis 2009 Update," EPRI, Palo Alto, CA: August 2009. 1019563.

"The Role of Electricity: A New Path to Secure, Competitive Energy in a Carbon-Constrained World," Eurelectric, Union of the Electricity Industry, Brussels, Belgium, March 2007.

"Energy Technology Perspectives 2010: Scenarios and Strategies to 2050," Presentation to the Press, Washington, DC, July 1, 2010, International Energy Agency (IEA)/.

Chapter 3

CO_2 Reductions through Expanded End-Use Applications of Electricity

INTRODUCTION

In most instances, applications of electricity are more efficient than their directly fossil-fueled counterpart, even taking into account losses from the electricity production, delivery, and conversion processes. For example, the electric heat pump, by pumping heat from the environment into a building, uses less fossil fuel than an efficient natural gas furnace. Likewise, freeze concentration in the food industry, by using a vapor compression system to freeze out water, uses less fossil fuel than the process of evaporating the water widely used today. In other cases, applications of electricity, such as the personal computer or fax machine, substantially reduce energy requirements and enhances productivity.

These are examples of beneficial electrification or beneficial new uses of electricity. While they increase electricity use from current levels, they reduce primary energy use, decrease CO_2 (and other emissions), and increase productivity.

Today, policymakers are facing a variety of issues in the energy area, including industrial productivity, national security, environmental concerns, limited resources, and increasing costs for resources. Many believe that a key path to follow is more efficient use of energy overall—not a reduction in electricity use.

Some argue that saving one unit of electricity is equivalent to preserving three units of primary energy. They believe that energy efficiency can be achieved by applying alternatives to electricity. Such perceptions have led many to believe that electricity use is inherently inefficient and that it adversely impacts the environment more than other energy forms. Some policymakers have mistakenly begun to encourage the use of fossil fuels in place of electricity and have argued that utilities should subsidize the conversion of electric-based end-use systems to natural gas.

While in some cases, electricity is a less efficient use of energy resources than direct fossil fuel consumption, in most cases, electricity is in fact a more efficient use of energy resources.

To determine true efficiency, the focus must be on total resource requirements, from the extraction of fossil fuel and other energy forms to the delivery of the end-use service. This means that when comparing the use of electricity to direct burning of a fossil fuel (natural gas or oil) for a given application, the entire system from raw resource through conversion, delivery, and end use must be considered. Because of the unique attributes of electricity, many new electric appliances and devices require less primary energy resources than comparable natural gas-fired or oil-fired systems. In these cases, not only are resources being used most efficiently, but the environmental impacts of fossil-fuel combustion are being minimized.

THE CLIMATE STABILIZATION CHALLENGE

Global climate change is one of the most complex environmental, energy, economic, and political issues confronting the world. Policies to address climate change are being considered at international, national, regional, state, and local levels. These policies include emissions limitations, investments in new technology, research to provide better understanding of climate science, demonstrations of green technology, and efforts to increase resilience to possible effects of climate change. The ultimate effectiveness of these policies and their cost will be determined by how they evolve into coordinated efforts over the coming decades that make the profound changes in the energy system, agricultural practices, and development patterns that would stabilize human impact on the world's climate. The challenge of stabilizing concentrations of greenhouse gases in the atmosphere while sustaining economic development is, at its heart, a technology challenge.

The earth's climate is governed by complex interactions among solar radiation, atmospheric processes, and surface effects of land and sea. In particular, increased concentration of heat-trapping "greenhouse gases" (GHGs) in the atmosphere has led to concerns that human activities could warm the earth and cause fundamental climate change. By far, the largest portion of the direct greenhouse effect is caused by carbon dioxide (CO_2), which is emitted by many human activities—particularly the burning of fossil fuels—and which can remain in the atmosphere for centuries (EPRI 1013041).

POWER DELIVERY AND END USE

Although reducing CO_2 emissions from power plants is generally considered the main contribution of the electric utility sector to stabilize atmospheric carbon levels; improving the efficiency of electricity end use and electrification can also make a vital contribution. This effort inherently involves making the power delivery infrastructure "smarter" in order to support various demand-side management (DSM) efforts. For example, real-time pricing and direct load-control initiatives have the potential to indirectly reduce CO_2 emissions by promoting improved end-use efficiency and providing utilities more flexibility in adjusting their generation mix in response to environmental and other needs. In such cases, a utility needs to be able to send a pricing or control signal to end-use equipment at a customer's site. The challenge is how to integrate advanced end-use technologies with a more responsive power delivery system so that customers can take advantage of emerging DSM opportunities. In addition, substantive reductions can be accomplished by substituting beneficial electric end uses for fossil-fueled end uses.

In 2004, EPRI took a major step toward facilitating such integration with the publication of the IntelliGrid[SM] Architecture—the first comprehensive technical framework for linking communications and electricity into a "smart grid" that will offer unprecedented flexibility and functionality required by an increasingly digital society. For consumers, a smart grid will enable introduction of advanced services, such as remote metering, outage detection, demand response, bill disaggregation, and real-time pricing—including the variable costs of CO_2 emissions control. For utilities, smart grid technologies will mean a more easily automated, self-healing power delivery system and an opportunity to reduce costs and develop new business enterprises. IntelliGrid can also help utility integration of distributed resources, including renewable energy, by coordinating the operation of these units with the grid at large.

The next step for IntelliGrid will be to develop a consumer portal that can provide the critical link between electricity customers and the integrated utility energy-communications network. Such a portal will enable the sharing of information between the network and the next generation of intelligent, highly efficient consumer appliances now entering the market. This two-way communications link will enable utilities to offer demand-response programs that operate at the level of individual appliances—improving load management while lowering customers' energy costs. EPRI is assuming an essential leadership role in this effort to ensure that the consumer portal uses open standards and provides "plug-and-play" capability for various communications media.

The use of electricity is at present a contributing factor to net carbon dioxide (CO$_2$) emissions. Growing concern over greenhouse gas emissions continues to focus resources to low-carbon power generation technologies to mitigate CO$_2$ emissions. Resources also are warranted for demand-side options, including necessary advances in the deployment and adoption of efficient electric end-use technologies.

There are two main mechanisms for reducing CO$_2$ emissions with electric end-use technologies: 1) upgrading existing electric technologies, processes, and building energy systems so as to increase end-use efficiency; and 2) expanding end-use applications of electricity. Upgrading existing electric end-use technologies embodies replacing or retrofitting older equipment with new, highly-efficient technologies. It also includes improving controls, operations, and maintenance, and reducing end-use energy needs by improving buildings and processes. In essence, this first mechanism is comprised of actions involving energy efficiency and demand response. The second mechanism, expanding end-use applications of electricity, involves replacing less efficient fossil-fueled technologies (existing or planned) with electric end-use technologies. It also encompasses deploying electric end-use technologies that result in overall energy, environmental, and economic benefits.

This chapter addresses the potential for expanding end-use applications of electricity to save energy and reduce CO$_2$ emissions. The focus is on converting residential, commercial, and industrial equipment and processes—existing or anticipated—from fossil-fueled end-use technologies to more efficient electric technologies.

TOTAL RESOURCE EFFICIENCY

To compare and analyze the total resource efficiencies of electric-based with fossil-fueled end uses, analysts consider the total energy "system" from the raw resource through conversion, delivery and utilization. In comparing fossil fuel with electricity use, analysts must consider all losses of conversion and delivery as well as the gains from leveraging alternative energy sources (e.g., a heat pump's use of solar heat or advanced technologies such as information technologies in homes.)

The system illustrated in Figure 3-1 depicts one unit of energy being derived from an oil or gas well. In an electrical system, it is converted to electricity at an efficiency ranging between 31% to 50%* (the lower range for coal—the higher for natural gas-fired units) and 100% (for nuclear, hydro and renewables). This conversion results in between 0.31 and 1.00 units of input energy electricity from the original 1.00 unit of fossil fuel. Applying a steady-state loss of 7% in electric transmission and distribution results in a delivery of between approximately .29 and .93 units of energy to the electric end use.

Typically, gas transmission and distribution systems lose 10% of their energy in pumping and leakage. This results in 0.90 units of en-

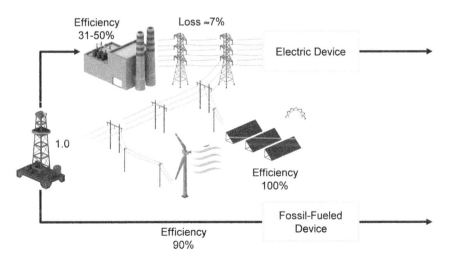

Figure 3-1. Illustration of Total System Efficiency of Electric vs. Gas

---

*Could be as high as 57%.

ergy being delivered to the gas end use from the original 1.00 available at the well head. This includes the losses in refining, transportation and dispensing.

Practical Examples

Several examples summarized below describe the benefits of examining total energy system efficiency; they reveal the breadth of the benefits of the wider use of electricity.

Freeze Concentration

Process industry opportunities to enhance electricity's value by increasing energy efficiency are truly promising: recent advances in freeze concentration provide one exciting example. Freeze concentration technologies separate substances in crystalline form at substantial savings in cost and energy. These technologies have a wide range of potential applications, from the preparation of food and chemicals to the treatment of wastewater. For example, the dairy industry is the largest user of energy for freeze concentration of any of the food industries. The equipment now used is generally antiquated—most typically employing evaporators. This technology is not nearly as efficient as freeze concentration and uses large supplies of fossil fuel. The dairy industry has exhibited interest in freeze concentration to replace thermal evaporation for several years.

As seen in Figure 3-2, freeze concentration uses an electric-based vapor compression system to freeze out the water in a product. Freezing requires a total of 144 Btu/lb of water, or 114 Btu/lb with heat recovery. Evaporating or boiling off water requires considerable more energy—in this case, a net of 700 Btu/lb with heat recovery. When the total systems are compared, the electric-based process has at least twice the overall efficiency as the natural gas-based process.

In addition, applying freeze concentration in the dairy industry not only yields superior quality products, but based on only 10% market penetration of this technology, the utilization of freeze concentration would save 3.4 x 1012 Btu/year of fossil fuel. This reduction in the use of fossil fuel would significantly reduce pollution and foreign oil dependence. In addition, freeze concentration technology operates at lower temperatures which reduces microbiological and enzymatic activity, thereby allowing better food quality, equipment utilization, and lower cleaning costs.

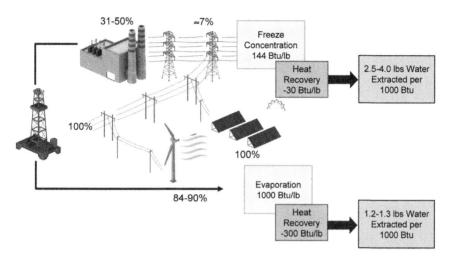

Figure 3-2. Illustration of Total System Efficiency of Freeze Concentration as Compared to Evaporation

Electric Heat Pump

One of the single greatest electric-powered technological advancements (outside of computers and medical electronics) is the electric heat pump. The most significant benefits of the heat pump are disguised by its very name "heat pump." It is fundamentally the most efficient solar machine available. The heat pump uses the heat in outside air or in the ground to transfer the sun's energy to buildings. This process yields tremendous system effectiveness. Electric heat pumps are available today with a steady-state efficiency almost four times that of an advanced gas-fired condensing furnace. Figure 3-3 depicts a heat pump unit in comparison with a gas-fired condensing furnace. Today's heat pumps incorporate electronic variable speed compressors and blower drives, as well as refrigerant circuits for year-round integrated water heating and defrosting methods.

Information Technologies

Increasingly, America's economy is becoming service-based. This shift has dramatically affected the use of information technologies in today's workplace. The growing need to manage and disseminate information has led to innumerable advances in electric-powered technologies. Concurrently, major changes in workers' job responsibilities have

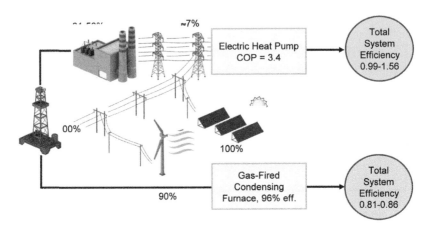

Figure 3-3. Illustration of Total System Efficiency of an Electric Heat Pump as Compared to a Gas-Fired Condensing Furnace

led to new stresses.

The need for psychological privacy to allow the eye and mind to identify a sense of private place may inhibit further dramatic increases in worker density in existing office environments. Some futurists have predicted that by the year 2030, we will see a move away from office-based employee teams toward the "electronic cottage." They envision the electronic cottage as a cozy room where all electronic gadgets would put the at-home office worker in kind of a command console—an intriguing idea that many workers are testing today. This "cottage" approach will not see significant use in this century, but may change energy use for these technologies by 2010.

Others have advanced the theory that the office of the future would contain fewer objects, in essence, be friendlier than a collection of all currently known office technologies. This theory could only be possible by integrating some technologies, perhaps combining the fax machine, telephone, photocopier and personal computer into one machine. Today's cutting-edge appliances are beginning to reveal this trend (for example, laser printers that now have fax and copy machine capabilities).

The movement of information presents the key to energy service needs in the commercial sector. Electrons will carry the information. These electrons, now normally supplied by electric utilities to just provide artificial illumination, motive power, cooling systems, and efficient heat pump systems, will increasingly energize information tech-

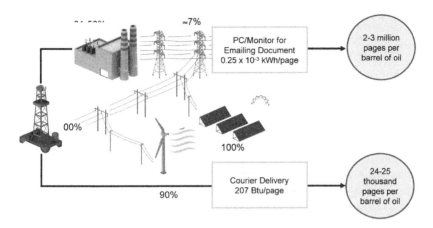

Figure 3-4. Illustration of Total System Efficiency of Information Transfer

nologies.

"Telecommuting," or the use of electronic systems instead of commuting, is increasing in popularity due to the growth in productivity which results. Use of computers represents one interesting example of the expanding, efficient use of electric-based technologies to move information. In the use of a computer and the Internet in relation to an express courier service, the Internet can send and deliver two to three million pages per barrel of oil as compared to 25,200 pages per barrel by a courier service.

Replacing fossil-fueled energy end uses with beneficial electric end-use technologies can deliver meaningful net reductions in CO$_2$ emissions. The most significant reductions can be achieved when implemented in conjunction with low-carbon generation, overall efficiency efforts, and when specific high-value replacement targets in energy end uses can be identified and acted upon.

SOURCES OF CO$_2$ REDUCTIONS

Beneficial new uses of electricity contribute to the reduction of CO$_2$ emissions in the following manner:

• A "beneficial" electric end-use technology replaces a fossil-fueled technology.

- The beneficial electric technology uses substantially less energy at the point of end use and often delivers superior quality and environmental performance.

- The beneficial electric technology has no CO_2 emissions and displaces a fossil-fueled device which had substantial CO_2 emissions at the point of end use.

- The electric technology causes some increase in CO_2 emissions in power generation. This is dependent on the generation mix, the pattern of end-use electric demand, and transmission and distribution losses.

- The net energy demand and net CO_2 emitted by electric generation from beneficial uses is substantially lower than the fossil-fueled technology due to the end-use efficiency of electricity.

- The fossil-fueled technology also has CO_2 and other emissions which are offset by the electric substitution. These may include energy used in drilling, refining, pumping, transportation, compression, and those due to spillage and leakage.

Replacing fossil-fueled energy end uses with beneficial electric end-use technologies can deliver meaningful net reductions in CO_2 emissions. The most significant reductions can be achieved when implemented in conjunction with low-carbon generation, overall efficiency efforts, and when specific high-value replacement targets in energy end uses can be identified and acted upon.

ENERGY SAVINGS FROM BENEFICIAL NEW USES

According to the Annual Energy Outlook (AEO) published by the U.S. Department of Energy's Energy Information Administration (EIA) in 2008, total annual energy consumption for the U.S. in the residential, commercial, industrial, and transportation sectors was 102.3 quadrillion Btus in 2008, including delivered energy and energy-related losses. EIA forecasts this consumption to increase by 15.3% to 118.0 quadrillion Btus in 2030, an annualized growth rate from 2008 to 2030 of 0.65%.

The AEO already accounts for market-driven efficiency improvements, the impacts of all currently legislated federal appliance standards

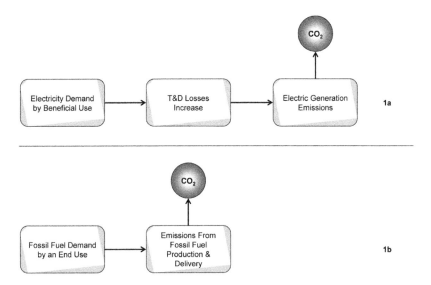

Figure 3-5. Comparative Illustrations of CO$_2$ Origins From Electric and Gas End Uses

and building codes (including the Energy Independence and Security Act of 2007) and rulemaking procedures. It is predicated on a relatively flat electricity price forecast in real dollars between 2008 and 2030. It also assumes continued contributions of existing utility- and government-sponsored end-use energy programs established prior to 2008.

Relative to the EIA AEO 2008 Reference Case, an EPRI study estimates (EPRI 1018906) between 1.71 and 5.32 quadrillion Btus per year of energy savings in 2030 due to expanded end-use applications of electricity. Expanded end-use applications of electricity have the potential to reduce the annual growth rate in energy consumption forecasted in EIA AEO 2008 between 2008 and 2030 of 0.65% by 10% to 32%, to an annual growth rate of 0.58% to 0.44%.

REDUCTIONS IN CO$_2$
EMISSIONS FROM BENEFICIAL NEW USES

According to EIA AEO 2008, annual energy-related CO$_2$ emissions for the U.S. in the residential, commercial, industrial, and transportation sectors is estimated at 5,983 million metric tons in 2008. EIA forecasts emissions to increase by 14.5% to 6,850 million metric tons in 2030, an annualized growth rate from 2008 to 2030 of 0.61%. This forecast is predicated on a relatively flat CO_2 intensity of the electricity generation mix between 2008 and 2030.

Relative to the EIA AEO 2008 Reference Case, the EPRI study identifies between 114 and 320 million metric tons per year of CO_2 emissions reductions in 2030 due to expanded end-use applications of electricity (EPRI 1018906). This demonstrates that expanded end-use applications of electricity have the potential to reduce the annual growth rate in CO_2 emissions forecasted in EIA AEO 2008 between 2008 and 2030 of 0.61% by 12% to 35%, to an annual growth rate of 0.54% to 0.40%.

The study used an analysis approach generally used in developing energy-efficiency and demand-response potential studies as well as in designing programs. The methodology uses a hybrid top-down and bottom-up approach for determining the potential for saving energy and reducing CO_2 emissions. For each sector, the baseline forecasts for energy use and CO_2 emissions were allocated down to fuel types, end uses, sub-sectors (for the industrial sector only), and technologies.

As stated earlier, the potential of an individual beneficial new use of electric end-use technology is a function of the opportunity's unit primary energy savings and reduction in CO_2 emissions relative to the fossil-fueled baseline technology. It is also a function of its technical applicability, the rate of adoption, and maximum market penetration values. For a given fuel type and end use, a baseline technology represents a discrete technology choice and is generally the most affordable. For example, natural gas furnaces are a baseline technology for process heating in the industrial sector. Efficient electric process heating technologies are applicable in existing industrial applications as replacements for natural gas furnaces that have reached the end of their expected useful life. They are also applicable to heating applications anticipated in the future to be met with natural gas furnaces.

The primary focus of the EPRI study was to determine the Technical Potential, which represents the maximum, technically-feasible impacts that would result if beneficial electric end-use technologies were to displace fossil-fueled technologies. This potential does not take into customer response.

The EPRI study made the very conservative assumption that new equipment is "phased-in" over time. Essentially, the phase-in potentials represent the energy savings and CO_2 reductions achieved if only the portion of the current stock of fossil-fueled equipment that has reached the end of its useful life and is due for turnover is replaced. As a result, the saturation of efficient electric end-use technologies is assumed to grow each year as more of the existing fossil-fueled equipment is up for replacement. In addition, any new equipment being placed in service in a given year due to market growth is assumed to be one of the applicable electric technologies.

The EPRI analysis provided energy and CO_2 impacts as a function of several parameters for each electric end-use technology including impacts by end-use sector (residential, commercial, and industrial) and impacts by displaced fuel type (natural gas, coal/coke, fuel oil, etc.).

IDENTIFYING AND SCREENING TECHNOLOGIES

Tables 3-1 through 3-6 summarize all of the electric end-use technologies which could be impacted by an expanded effort to apply beneficial new uses of electricity. The technologies are categorized by sector and applicable end-use area.

The technologies in the long list were screened to determine if they met the criteria required for inclusion in the final study. The primary criterion of importance was the ability of the electric end-use technology to decrease overall CO_2 emissions relative to fossil-fueled technologies. If displacing a fossil-fueled technology with a given electric technology increased overall CO_2 emissions during the study period, the technology "failed" the screening process and was eliminated from the calculations. Only the short-list of "passing"—or favorable—electric end-use technologies was included in the estimates. Note that as the carbon intensity reduces over time, many—if not all—of these technologies will become advantageous in reducing net CO_2. In fact, the outcome of the screening exercise is highly dependent on the projected CO_2 intensity of the electricity generation mix of the future. The baseline forecast used for the EPRI study was derived from the EIA AEO 2008. This forecast information does not presume that future activities may lower the CO_2 intensity of the generation. With lower carbon emissions, more of the electric technologies analyzed would fall under the favorable category.

In fact, new electric technologies will become available which will become favorable if the CO_2 emissions factors decreased even modestly between now and 2030. A 10% reduction in CO_2 emissions factors would

Table 3-1. Efficient Electric End-Use Technologies Analyzed for Potential Displacement of Fossil-Fueled Technologies in the Residential Sector (Source: EPRI 1018906)

End-Use Area	Efficient Electric End-Use Technology	Displaced Fossil-Fueled Technology
Clothes Drying	Heat Pump Clothes Dryer	Natural Gas Clothes Dryer
Cooking	Electric Oven	Natural Gas Oven
	Inductive Range Top	Natural Gas Range Top
Pool/Spa Heating	Heat Pump Pool/Spa Heater	Distillate Fuel Oil Pool Heater
		Natural Gas Pool Heater
		Propane/LPG Pool Heater
Space Cooling	Air-Source Heat Pump	Natural Gas Heat Pump
	Ground-Source Heat Pump	Natural Gas Heat Pump
Space Heating	Air-Source Heat Pump	Coal Heating Stove
		Distillate Fuel Oil Boiler
		Distillate Fuel Oil Furnace
		Kerosene Furnace
		Kerosene Portable Heater
		Natural Gas Boiler
		Natural Gas Furnace
		Natural Gas Heat Pump
		Natural Gas Vented Direct Heating
		Propane/LPG Boiler
		Propane/LPG Furnace
		Propane/LPG Vented Direct Heating
	Ground-Source Heat Pump	*Same as air-source heat pump*
Water Heating	Electric Instantaneous Water Heater	Distillate Fuel Oil Storage
		Natural Gas Instantaneous
		Natural Gas Storage
		Propane/LPG Instantaneous
		Propane/LPG Storage
	Heat Pump Water Heater	*Same as electric instantaneous*

make the following industrial electric end-use technologies favorable in the regions indicated:

- Northeast
  - Electric boilers replacing coal-fired boilers;
  - Induction heating replacing direct-fired natural gas;
  - Radio frequency heating replacing direct-fired natural gas;

Table 3-2. Efficient Electric End-Use Technologies Analyzed for Potential Displacement of Fossil-Fueled Technologies in the Commercial Sector (Source: EPRI 1018906)

End-Use Area	Efficient Electric End-Use Technology	Displaced Fossil-Fueled Technology
Clothes Drying	Heat Pump Clothes Dryer	Natural Gas Clothes Dryer
Cooking	Electric Braising Pan	Natural Gas Braising Pan
	Electric Broiler	Natural Gas Broiler
	Electric Griddle	Natural Gas Griddle
	Electric Fryer, Flat Bottom	Natural Gas Fryer, Flat Bottom
	Electric Fryer, Open Deep Fat	Natural Gas Fryer, Open Deep Fat
	Electric Fryer, Pressure/Kettle	Natural Gas Fryer, Pressure/Kettle
	Electric Oven, Conveyor	Natural Gas Oven, Conveyor
	Electric Oven, Deck	Natural Gas Oven, Deck
	Electric Oven, Rotisserie	Natural Gas Oven, Rotisserie
	Electric Oven, Standard/Convection	Natural Gas Oven, Standard/Convection
	Electric Range Top	Natural Gas Range Top
	Electric Steamer, Compartment	Natural Gas Steamer, Compartment
	Electric Steamer, Kettle	Natural Gas Steamer, Kettle
	Electric Wok	Natural Gas Wok
Pool/Spa Heating	Heat Pump Pool/Spa Heater	Distillate Fuel Oil Pool Heater
		Natural Gas Pool Heater
Space Cooling	Air-Source Heat Pump	Natural Gas Space Cooling
	Ground-Source Heat Pump	Natural Gas Space Cooling
Space Heating	Electric Boiler	Distillate Fuel Oil Boiler
		Natural Gas Boiler
		Residual Fuel Oil Boiler
	Air-Source Heat Pump	Distillate Fuel Oil Boiler
		Distillate Fuel Oil Furnace
		Natural Gas Boiler
		Natural Gas Furnace
		Natural Gas Vented Direct Heating
		Residual Fuel Oil Boiler
	Ground-Source Heat Pump	Same as air-source heat pump
Water Heating	Heat Pump Water Heater	Distillate Fuel Oil Storage
		Natural Gas Instantaneous
		Natural Gas Storage

End-Use Area	Efficient Electric End-Use Technology	Displaced Fossil-Fueled Technology
Manufacturing Industries		
Boilers	Electric Boiler	Natural Gas Boiler
		Fuel Oil Boiler
		Coal-Fired Boiler
	Electric Drive	Natural Gas Boiler
		Fuel Oil Boiler
		Coal-Fired Boiler
Space Heating	Heat Pump	Natural Gas Furnace
Process Heating	Heat Pump	Natural Gas Furnace
	Induction Heating	Direct-Fired Natural Gas
	Radio Frequency Heating	Direct-Fired Natural Gas
	Microwave Heating	Direct-Fired Natural Gas
	Electric Infrared Heating	Direct-Fired Natural Gas
	UV Heating	Direct-Fired Natural Gas
	Electric Arc Furnace	Coke Blast Furnace
	Electric Induction Melting	Natural Gas Furnace
	Plasma Melting	Natural Gas Furnace
	Electrolytic Reduction	Natural Gas Furnace

Table 3-3. Efficient Electric End-Use Technologies Analyzed for Potential Displacement of Fossil-Fueled Technologies in the Industrial Sector (Source: EPRI 1018906)

Electric End-Use Technologies Yielding Net Reductions in CO_2 Emissions		
Electric Technologies Analyzed (Long List)	Net Decrease in CO_2 Emissions?	Favorable Electric Technologies (Short List)
Heat Pump Clothes Dryer	Yes (All but Midwest)	Heat Pump Clothes Dryer
Heat Pump Pool/Spa Heater	Yes	Heat Pump Pool/Spa Heater
Air-Source Heat Pump, Cooling	Yes	Air-Source Heat Pump, Cooling
Air-Source Heat Pump, Heating	Yes	Air-Source Heat Pump, Heating
Ground-Source Heat Pump, Cooling	Yes	Ground-Source Heat Pump, Cooling
Ground-Source Heat Pump, Heating	Yes	Ground-Source Heat Pump, Heating
Electric Instantaneous Water Heater	Yes (Only Northeast)	Electric Instantaneous Water Heater
Heat Pump Water Heater	Yes	Heat Pump Water Heater
Electric Convection Oven	No	
Electric Induction Range Top	No	

Table 3-4. Short List of Favorable Residential Electric End-Use Technologies Capable of Yielding Net Reductions in CO_2 Emissions (Source: EPRI 1018906)

Electric End-Use Technologies Yielding Net Reductions in CO_2 Emissions		
Electric Technologies Analyzed (Long List)	Net Decrease in CO_2 Emissions?	Favorable Electric Technologies (Short List)
Heat Pump Clothes Dryers	Yes (All but Midwest)	Heat Pump Clothes Dryers
Electric Broiler	Yes (Northeast, West)	Electric Broiler
Electric Oven, Conveyor	Yes (Northeast, West)	Electric Oven, Conveyor
Electric Oven, Deck	Yes (Northeast)	Electric Oven, Deck
Electric Oven, Rotisserie	Yes (Northeast)	Electric Oven, Rotisserie
Electric Wok	Yes (Northeast, West)	Electric Wok
Heat Pump Pool/Spa Heater	Yes	Heat Pump Pool/Spa Heater
Air-Source Heat Pump, Cooling	Yes	Air-Source Heat Pump, Cooling
Air-Source Heat Pump, Heating	Yes	Air-Source Heat Pump, Heating
Ground-Source Heat Pump, Cooling	Yes	Ground-Source Heat Pump, Cooling
Ground -Source Heat Pump, Heating	Yes	Ground -Source Heat Pump, Heating
Heat Pump Water Heater	Yes	Heat Pump Water Heater

Table 3-5. Short List of Favorable Commercial Electric End-Use Technologies Capable of Yielding Net Reductions in CO_2 Emissions (Source: EPRI 1018906)

- Microwave heating replacing direct-fired natural gas;
- Electric infrared heating replacing direct-fired natural gas; and
- UV heating replacing direct-fired natural gas.

- Midwest
  - Plasma melting replacing natural gas furnaces; and
  - Electric induction melting replacing natural gas furnaces.

TECHNICAL AND REALISTIC POTENTIALS BY SECTOR

Table 3-7 summarizes the technical results of the EPRI study by end-use sector for the beneficial electric end-use technologies. The table shows that the residential sector has the greatest promise for beneficial impacts. The commercial and industrial sectors follow with values that are roughly comparable to each other. The Technical Potential impacts of all three sectors combined are energy savings of 5.32 quadrillion Btus per year and CO_2 emissions reductions of 320 million metric tons per year in 2030 relative to the baseline forecast.

Table 3-6. Short List of Favorable Industrial Electric End-Use Technologies Capable of Yielding Net Reductions in CO_2 Emissions (Source: EPRI 1018906)

Electric End-Use Technologies Yielding Net Reductions in CO_2 Emissions		
Electric Technologies Analyzed (Long List)	Net Decrease in CO_2 Emissions?	Favorable Electric Technologies (Short List)
Heat Pump	Yes	Heat Pump
Plasma Melting	Yes	Plasma Melting
Electrolytic Reduction	Yes	Electrolytic Reduction
Electric Induction Melting	Yes	Electric Induction Melting
Electric Arc Furnace	Yes	Electric Arc Furnace

Table 3-7. Technical and Realistic Potential: Annual Impacts on Primary Energy Use and CO_2 Emissions by Sector (Source: EPRI 1018906)

Baseline	EIA AEO 2008 Baseline Forecast of Primary Energy Use (Quadrillion BTUs per Year)			EIA AEO 2008 Baseline Forecast of Energy-Related CO_2 Emissions (Million Metric Tons per Year)		
Sector	2010	2020	2030	2010	2020	2030
Residential	22.2	23.4	25.0	1,259	1,323	1,450
Commercial	18.7	22.0	25.0	1,080	1,265	1,474
Industrial	33.3	34.3	35.0	1,693	1,718	1,733
Transportation	29.0	31.2	33.0	1,980	2,077	2,193
U.S.	103.3	110.8	118.0	6,012	6,383	6,850
Technical Potential	Decrease in Primary Energy Use (Quadrillion BTUs per Year)			Decrease in CO_2 Emissions (Million Metric Tons per Year)		
Sector	2010	2020	2030	2010	2020	2030
Residential	0.352	2.20	2.96	21.7	135	178
Commercial	0.118	0.72	1.09	7.75	47.2	69.1
Industrial	0.123	0.72	1.27	7.30	43.7	73.1
U.S.	0.593	3.64	5.32	36.7	226	320
Realistic Potential	Decrease in Primary Energy Use (Quadrillion BTUs per Year)			Decrease in CO_2 Emissions (Million Metric Tons per Year)		
Sector	2010	2020	2030	2010	2020	2030
Residential	0.055	0.417	0.633	4.97	34.3	47.3
Commercial	0.018	0.146	0.277	1.93	13.3	21.4
Industrial	0.078	0.457	0.802	4.57	27.1	45.1
U.S.	0.152	1.02	1.71	11.5	74.7	114

In terms of the potential for energy savings, the Technical Potential is associated with a 4.5% reduction relative to the baseline in the year 2030. For CO_2 emissions, the technical potential reduces baseline emissions by 4.7% in the year 2030.

Figures 3-6 and 3-7 graphically depict the impacts of the electric end-use technologies on primary energy use and energy-related CO_2 emissions, respectively. The values are expressed in terms of the cumulative Technical Potential impacts between 2009 and 2030. In all three sectors, heat pumps are the technology with the greatest promise for saving energy and reducing CO_2 emissions. In the industrial sector, electric arc furnaces have a significant potential for beneficial impacts as well. In addition, electrolytic reduction, electric induction melting, and plasma melting also show promise. Under a less carbon-intensive future generation mix, more technologies would cross the line and become favorable in regards to saving energy and reducing emissions.

In both the residential and commercial sectors, the end-use areas with the most potential for beneficial impacts are space heating and then

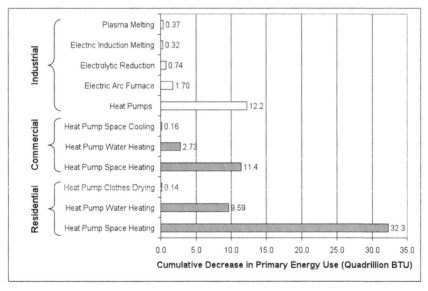

Figure 3-6. Technical Potential: Cumulative Decrease in Primary Energy Use Between 2009 and 2030 by Sector and Efficient Electric End-Use Technology (Source: EPRI 1018906)

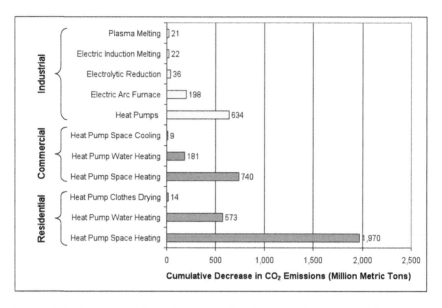

Figure 3-7. Technical Potential: Cumulative Decrease in Energy-Related CO_2 Emissions between 2009 and 2030 by Sector and Efficient Electric End-Use Technology (Source: EPRI 1018906)

water heating. Clothes drying (residential) and space cooling (commercial) also exhibit potential. In the industrial sector, process heating is the predominant end-use area showing potential, followed by space heating.

JAPANESE STUDY

Japan has adopted an aggressive strategy to promote electrification through the adoption of heat pumps. The Japanese see heat pumps as an effective environmental tool that has come to the fore in Japan and can be implemented by households and companies without the need for any major changes to their current way of life.

In the Japanese residential sector, around 58% of energy demand is for air conditioning needs and hot water supplies. Kitchen use accounts for around 7%, and demand for other domestic appliance products accounts for some 35%. Similarly in the commercial sector, heating and

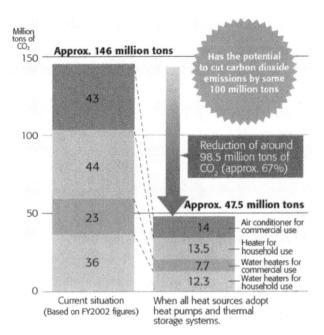

Figure 3-8. Japanese Heat Pump Study (Source: Science, 2006)

cooling applications and hot water supply account for more than 50% of total demand. For the most part, this demand is still being met through the use of the heat generated through the burning of fossil fuels.

A Japanese study investigated what would happen if heat pumps were introduced into all household and commercial air conditioning and hot water systems. Currently, the household and commercial sectors combined are responsible for some 150 million tons of CO_2 emissions annually in Japan. If heat pumps were introduced across the board, then this could reduce CO_2 emissions by households by 54.2 million tons and by the commercial sector by 44.3 tons, reducing CO_2 emissions overall by a total of 98.5 million tons (Science, 2006)

EUROPEAN STUDY

Figure 3-9 is from Eurelectric, the Union of the European Electric Industry, and endorsed by the European Union. The project was entitled "The Role of Electricity" and studied the potential to reduce CO_2 emissions. This study finds an even more drastic potential for CO_2 reduction than the Japanese study. It makes aggressive assumptions about elec-

Role of Electricity Scenario – EU 25 CO_2 Emissions

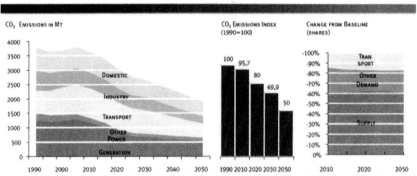

Figure 3-9. Eurelectric Study (Source: Eurelectric, 2007)

trification, as well as unleashing the potential of energy efficiency and increased use of low-carbon generation. It provides further evidence of the world-wide potential for electrification.

UNITED KINGDOM

The United Kingdom's Energy Research Centre (UKERC, 2009) studied a variety of energy future scenarios. These scenarios were analyzed in order to respond to the UK government's target of reducing CO_2 emissions by 80% below 1990 levels by 2050. Among the conclusions drawn by this study was that the electricity system must effectively be de-carbonized and that fossil fuel usage must be electrified in transportation and in heating. The principle transportation technology identified were personal electric vehicles. The principle heating technologies involved were both air source and ground source heat pumps. One specific conclusion was that the market penetration of heat pumps reaches between 10% and 60% by 2050.

CONCLUSION

While there are some cases where electricity may be less efficient in the use of energy resources, it is increasing obvious that many end uses of electricity are superior—even with today's generation mix. And as that generation mix increasingly de-carbonizes—electricity will enable society to greatly reduce CO_2 emissions.

References

"The Potential to Reduce CO_2 Emissions by Expanding End-Use Applications of Electricity, Executive Summary," C.W. Gellings, EPRI, Palo Alto, CA: 2009. 1018906.

"Program on Technology Innovation: Electric Technology in a Carbon-Constrained World," T. Wilson, EPRI, Palo Alto, CA: 2005. 1013041.

"Climate Contingency Roadmap: The U.S. Electric Sector and Climate Change," T. Wilson, EPRI, Palo Alto, CA: 2003. 1009311.

"Beneficial Electrification: An Assessment of Technical Potential," P. Hummel, EPRI, Palo Alto, CA: 1992. CU-7441.

"Heat Pumps the Trump Card in Fight Against Global Warming," Y. Takashi, *Science—The Japan Journal*, October 2006.

"Making the Transition to a Secure and Low-Carbon Energy System: Synthesis Report," UK Energy Research Centre (UKERC), 2009.

Chapter 4

Electric On-Road Transportation

The concept of using electricity to power trucks, trains and cars has been around since the late 1860s. Early applications were in light rail, predominantly street cars. These applications eventually spread to electrifying trains. Trains throughout the world are increasingly powered by electricity. Some of the first automobiles were electric. Unfortunately, at that time, battery technology limited the range of electric vehicles, and they never developed substantial market share. Only recently has this application re-emerged.

Applying electricity had immediate appeal since it is clean at the point of end use. Electric motors are highly efficient, offer ease of control, phenomenal torque, and they are lighter weight (hp/lb) than internal combustion engines. In addition, in transportation applications, they can generate electricity during braking—called regenerative braking.

Somewhere between 1832 and 1839, Scottish inventor, Robert Anderson, built a very simple electric carriage. Meanwhile, in the U.S., Thomas Davenport built an electric road vehicle in 1835. These first vehicles were largely experimental.

Rechargeable batteries were not available until later in the 19th century, with the first commercially available rechargeable batteries becoming available in 1881. The first production electric car was built in London in 1884 by Thomas Parker, a British inventor who was responsible for electrifying the London Underground (Boxwell, 2010).

The first successful electric automobile in the United States was built in 1891 by William Morrison of Des Moines, Iowa. The vehicle was the only American automobile exhibited along with five European vehicles at the 1893 World's Columbian Exhibition in Chicago. Morrison's automobile was equipped with a four-horsepower motor and a 24-cell battery weighing 768 pounds—more than half of the vehicle's total weight. Top speed was a breathtaking (for the time) 14 miles per hour. In 1897, the Pope Manufacturing Company of Hartford, Connecticut, became the first American builder and seller of electric cars in quantity. Pope's Columbia Electric Phaeton, Mark III, weighed some 1,800 pounds, including 850 pounds of batteries, and had four forward speeds of 3, 6, 12 and 15 miles per hour (Sulzberger, 2004).

These first electric cars performed well and were reliable—albeit with limited range. Internal combustion engine-powered cars were very noisy, emitted fumes and smoke, and were unreliable. In addition, these early cars had to be hand cranked to be started. Hand cranking results in many injuries.

By the end of the First World War, gasoline cars were vastly more reliable, petroleum was significantly cheaper and more easily available, and gasoline-powered cars were far cheaper than electric cars. Service stations also were appearing, making it easier to buy fuel.

Only more recently has an additional attribute of electricity use in transportation become obvious: The fact that electric transportation systems use far less net energy and have far lower net CO$_2$ (and other) emissions than their fossil-fueled counterparts. The reason for this has to do with the total fuel cycle.

Fossil-fueled transportation systems typically use internal combustion (IC) engines. IC engines vary inefficiency according to the application—but overall efficiencies of 25 to 40% are typical. This means that 60 to 75% of the fossil fuel delivered to the IC engine is rejected as waste heat or pollutants or lost in braking. Another substantial loss in the IC engine-driven transportation system is in extracting, refining and transporting the fossil fuel. The losses in this part of the value chain can range from 10% (for natural gas) to 24% (for diesel fuel or gasoline).

Electric transportation systems are very efficient at the point of end use with efficiencies of between 80 and 95%, if properly engineered. However, there are substantial losses in the production and delivery of electricity. Efficiency of generation vary according to the technology. The lowest efficiencies are from old coal plants (perhaps 31%) to modern gas-fired combined-cycle plants (approximately 50 to 57%) to hydro, wind, solar and nuclear (at 100%). Losses in delivery or losses in the electric transmission and distribution system average 7% in the U.S.

The net losses in the electricity system and its overall efficiency as compared to the fossil-fueled system is dependent to a significant extent on the generation mix. The electricity sector continues to expand the use of gas-fired combined-cycle combustion turbines, wind, solar and nuclear power generation. Because of the current and increasing use of more efficient means of generating electricity, the electric transportation system is now—and continues to be—more efficient and less polluting than fossil-fueled transportation.

The following assessment of plug-in hybrid electric vehicles offers an example of the comparative advantages of electricity.

Figure 4-1 illustrates a "well-to-wheels" analysis of an electric vehicle energy consumption as compared to a gasoline-powered internal combustion (IC) engine vehicle. Assuming electric generation efficiency of between 31 and 50%, T&D losses of 7% and electric vehicle charging efficiency of 86% and 0.2 KWh per mile: the electric vehicle operates at an equivalent of between 45 and 72 miles per gallon (mpg). The reader should note that as increasing amounts of nuclear and renewable power production are added to the system at 100% efficiency, the mpg equivalent of the electric vehicle will increase dramatically.

In contrast, the gasoline-powered vehicle with a generous on-board fuel economy of 23 mpg will have a net efficiency of 25 to 27 mpg. In practice, this is less than one-half the electric fuel economy, and if the actual generation mix of most utilities was used in the analysis—less than one-third.

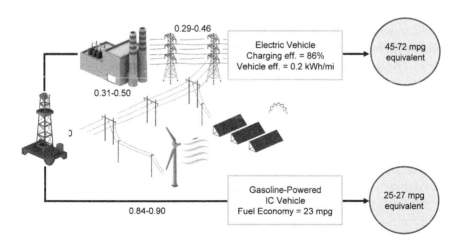

Figure 4-1. Illustration of Total System Efficiency of an Electric Vehicle Compared to a Gasoline Vehicle

THE REBIRTH OF ELECTRIC CARS

From the first world war until the 1990s, electric cars were a curiosity. There were small niche markets such as delivery vans in the U.K. and experimental "concept" cars developed by U.S. automakers. Air pollution and climate change concepts in the 1990s rejuvenated interest in electric cars.

Today, the family of electric cars includes hybrids, plug-in hybrid electric vehicles (PHEVs), electric vehicles (EVs), and extended-range electric vehicles. All these plug-in vehicles are today referred to as plug-in electric vehicles (PEVs).

The California Air Resources Board (CARB) passed a ruling called the Low Emission Vehicle Program in 1990 to promote the use of zero-emission vehicles (ZEVs). The law stated that 2% of all new vehicles sold in California were to be "so-called" zero-emission vehicles by 1998, rising to 10% of all new vehicles by 2003.

U.S. automakers developed electric vehicles in order to attempt to comply with the new law. General Motors launched the "Impact" electric sedan—later renamed the EV1. (GM's marketers suggested that naming a car the Impact was unwise.) Ford launched an electric version of its Explorer SUV, Chrysler bought an electric car maker, Toyota produced an electric RAV-4, and Honda produced a city car.

The EV1 was the most successful of these early attempts. Even though it had a host of technical problems. However, its owners were enthusiastic about the vehicles. In general, these early vehicles were very expensive, due primarily to battery limits, and had modest range.

The game accelerated in 2001 when Toyota introduced a hybrid electric vehicle, called the Prius sedan to the U.S. market. A five-seat compact with a 1.5 liter four-cylinder gasoline engine and an electric motor powered by a 276-volt NiMH battery pack, the Prius was a practical car for the average American. It was followed with the introduction of the second-generation model, featuring a five-door hatchback body, in the 2004 model year. A sales success, the 2004 to 2009 Prius is the most recognized hybrid on the market.

In addition, early 2001 saw a leap in battery development with the lithium-ion concept. Being smaller and more durable than nickel-metal batteries, this battery showed more promise for bringing the hybrids and all electrics greater range and power.

The Hybrid Story

The obvious solution to the limited range of these 1990s electric vehicles was to add a gasoline or diesel engine to the electric vehicle, either as a motor generator or as a second in-line power plant. This concept was openly discussed and debated—but offered such a non-traditional approach that the automakers feigned no interest.

With the right cost and performance, the hybrid could offer some of the benefits of electric transportation without the limited range of a straight electric vehicle. An added benefit of hybrids is the ability to recover some of the braking energy lost in internal combustion (IC) engine-powered vehicles by regenerative braking.

The Prius Story

In 1993, the Clinton Administration formed the Partnership for the Next Generation of Vehicles (PNGV). The project was designed to fund research into developing family-sized vehicles that could deliver 80 miles per gallon. Toyota was denied the opportunity to join since they were a Japanese automotive company. Toyota was reportedly very upset by not being included and secretly began a project of its own (John's Stuff, 2008).

Toyota's efforts became visible in October 1995 when they revealed the first Prius as a concept car. It used a new propulsion system Toyota had created called EMS which stands for Energy Management System. That design consisted of an electric motor connected to a regular gasoline engine. In 1997, Toyota revealed the production version of Prius exclaiming it would be available for purchase in just three months. The Prius was a real production vehicle—the first in the world. Reportedly, this "Absolutely horrified the American automotive companies" (John's Stuff, 2008). The the U.S. automakers didn't even have prototypes developed yet. This model used the THS (Toyota Hybrid System) a design advanced beyond EMS.

In December 1997, sales of the "original" generation of Prius began in Japan and continued there through 2001. By 2001, the Prius sported a 76 hp, 57 kW power trains using a 500-volt hybrid system with a touch-sensitive multi-display and electric air conditioning.

In December 1999, Toyota announced the next generation of Prius would be available for sale on the American market in the second half of 2000 and began displaying a 2000 model at auto shows throughout the United States.

In January 2000, the American automotive companies demonstrated their PNGV prototypes. Hybrids were becoming popular.

In June 2000, Toyota debuted a purchase web site where those interested in purchasing a Prius could begin the process by filling out an on-line request form. This "pioneer purchase" method was the only way to initially get a Prius. By late summer, delivery began. This resulted in a massive backorder delay due to an overwhelming surge in consumer demand.

Until the Prius, the U.S. auto manufacturers dogmatically insisted that they would never manufacture vehicles with what they referred to as "two fuel sources" on board. With Toyota's success—the race was on! The market success of the Toyota Prius combined with higher fuel prices (above $4.00 per gallon in 2008) across the U.S., prompted an explosion of hybrid cars throughout the industry.

For example, in 2007, For announced the 2008 Mercury Mariner Hybrid featured regenerative braking and was ranked a competitor to the Toyota Prius and Honda Civic in consumer comparisons for safety and fuel mileage (Anderson, 2009).

The Toyota Prius Hybrid, with its recognizable design, enjoyed record sales. The first six months of 2007 showed an increase in sales to 94,500 cars, up over 90% from the previous year.

In December 2007, the Electric Vehicle Symposium, held in Anaheim, California, showcased many types of electric vehicles including scooters, bicycles, skateboards, mini-cars, and even trucks.

Further hybrid developments included:

- 2007: GM Saturn launches the VUE.
- 2007: GM Hybrid "2 Mode."
- 2007: GM Yukon and Chevy Tahoe Hybrid SUVs.
- 2007: Toyota SUV Highlander Hybrid.
- 2007: Ford Escape Hybrid launch expanded.
- 2007: Ford Mercury Mariner launch expanded.
- 2009: GM Silverado and GMC Sierra Hybrid pick-ups.

Other PEV Developments
Tesla

Tesla Motors was founded in 2003 by a group of Silicon Valley engineers who set out to prove that electric vehicles could be commercially viable. The Tesla all-electric two-seat roadster hit the streets in early 2008. Two years later over 1000 roadsters are in operation in more than 25 countries (http://www.teslamotors.com/about). It has a carbon-fiber body with a top speed of 130 mph and a range of 250 miles. Tesla's battery packs and power trains were developed in cooperation with other car manufacturers.

Chevrolet Volt

The 2011 Chevrolet Volt is Chevy's much-anticipated new extended-range electric vehicle. It is an electrically driven hybrid, featuring an electric-only mode with a range of up to 40 miles. It will probably be the most fuel-efficient car on the market.

The Volt is a four-seat, four-door "series plug-in hybrid" hatchback with a lithium-ion battery pack that can power the car's 149-horsepower (11-kilowatt) electric motor by itself for an estimated 40 miles in the city. After that, the gasoline-powered in-line four-cylinder engine is used to supply electricity to the motor for as many as 300 additional miles.

Standard features include 17-inch alloy wheels, automatic headlights, heated mirrors, keyless ignition, remote ignition, automatic climate control, cruise control, auto-dimming rearview mirror, six-way manual front seats, tilt-and-telescoping steering wheel, cloth upholstery, Bluetooth, OnStar, a navigation system with touch screen, voice controls and real-time traffic, and a six-speaker Bose stereo with CD/DVD player, auxiliary audio jack, iPod/USB interface and 30GB of digital music storage.

The Premium Trim package adds leather upholstery, a leather-wrapped steering wheel, and heated front seats. The Rear Camera and Park Assist packages add a rearview camera and front and rear parking sensors.

The 2011 Volt is a front-wheel-drive vehicle powered by an electric motor rated at 149 horsepower (111 kilowatts) and 273 pound-feet of torque. This motor draws power from a lithium-ion battery pack until the battery charge is 70% depleted. At that point, the Volt's 1.4 liter four-cylinder internal combustion engine, which runs on either gasoline or E85, comes to life as a replacement power source for the electric motor. The battery can only be completely recharged through either a 120-volt or 240-volt outlet, but regenerative braking and the engine generator can replenish it slightly. There is no transmission in the traditional sense; rather, the Volt employs a single reduction gear to send the electric motor's power to the front wheels. General Motors claims the Volt can sprint to 60 miles per hour in 8.5 to 9 seconds with full battery power.

Nissan Leaf

Nissan has announced its intention to sell its "Leaf" in December 2010. The Leaf is an all-electric car with a range of 100 miles. The announced price starts at $32,780 in the U.S.

The Leaf uses a front-mounted electric motor driving the wheels, powered by a 24 kWh/90 kW lithium ion battery pack. The battery pack is made of air-cooled stacked laminar cells with lithium manganate in the cathode. The battery and control module together weigh 300 kilograms (660 pounds). The battery is expected to have 70 to 80% of capacity after 10 years, depending on how much (400-volt) fast charging is done, and also average ambient temperature. Nissan states the battery will have a "lifespan of 5 to 10 years under normal use." To keep the center of gravity as low as possible, the battery is housed partly below the front seats, in a thin layer below the rear floor, but mostly in a long rack below the rear seats. The battery actually consists of 48 modules with each module containing 4 cells (Wikipedia, 2010).

Using a trickle charge cable provided by Nissan, the Leaf can be charged in about 20 hours from a standard 120-volt, 20-amp outlet. It can be charged in 8 hours from a 240-volt supply depending on amperage. Using level 3 quick charging, it can be charged to 80% capacity in about 30 minutes.

The Leaf's frontal styling is characterized by a sharp V-shape design with large, up-slanting light-emitting diode (LED) headlights that create a distinctive blue internal design. The headlights consume 50% less electricity than halogen lamps.

Nissan says that the car has a top speed of over 87 mph using a motor rated at 80 kW (110 hp).

Renault-Nissan

In addition to the influence of the U.S. and Japanese automakers on world-wide interest in hybrid, plug-in hybrid, extended-range electric vehicles, and pure electric vehicles, some countries have launched special efforts. For example, in France, Renault and Nissan have formed an alliance in which they aim to become a leader in electric vehicles. The French estimate that automobiles account for 12% of the planet's

CO_2 emissions and 25% of world oil consumption. Because of France's low-carbon electricity production process, electric vehicles use only 12 g/km "well to wheel" (average emissions linked to the production of electricity) as compared with 105 g/km for the best hybrid and 137 g/km for the best diesel-powered vehicle of equivalent size (www.renault-ze.com).

The Renault-Nissan Alliance plans to market four "zero-emission (Z.E.) electric vehicles in order to meet a range of customer needs and make innovation accessible. These are as follows.

- **Twizy Z.E. Concept**—A tandem two-seater with four wheels targeting urbanites looking for an alternative to a moped that is safer, more comfortable, and more environmental. Its motor has a 15 kW (20 hp), 70 Nm torque engine with a top speed of 70 kph, and a 100 km range.

- **Zoe Z.E. Concept**—A versatile city car that is 4.10 m long for all day-to-day uses. Its motor has a 70 kW (95 hp), 225 Nm torque engine with a top speed of 130 kph and a rang of 160 km.

- **Fluence A.E. Concept**—This comfortable five-seater sedan is for commuting and family weekend trips. It has a 70 kW (95 ph), 225 Nm torque engine with a top speed of 130 kph and 160 km range.

- **Kangoo Z.E. Concept**—This is a light commercial vehicle for urban deliveries. It has a 70 kW (95 hp), 226 Nm torque engine and a top speed of 130 kph and a range of 160 km.

ENVIRONMENTAL ASSESSMENT OF PLUG-IN
HYBRID ELECTRIC VEHICLES

One comprehensive environmental assessment of electric transportation was made by the Electric Power Research Institute (EPRI) and the National Resources Defense Council (NRDC) to examine the greenhouse gas (GHG) emissions and air quality impacts of plug-in hybrid electric vehicles (PHEV). The purpose of the study was to evaluate the nationwide environmental impacts of potentially large numbers of PHEVs over the time period of 2010 to 2050 (EPRI 1015325).

The objectives of the EPRI-NRDC study were the following:

- Understand the impact of widespread PHEV adoption on full fuel-cycle greenhouse gas emissions.

- Model the impact of a high level of PHEV adoption on air quality.

- Develop a consistent analysis methodology for determination of the environmental impact of future vehicle technology and electric sector scenarios.

The EPRI-NRDC study is a "well-to-wheels" analysis which accounted for emissions from the generation of electricity to charge PHEV batteries and from the production, distribution and consumption of gasoline and diesel motor fuels.

The researchers used detailed models of the U.S. electric and transportation sectors and created a series of scenarios to examine the assumed changes in both sectors over the 2010 to 2050 timeframe of the study.

- Three scenarios were used representing high, medium, and low levels of both CO_2 and total GHG emissions intensity for the electric sector as determined by the mix of generating technologies and other factors. While CO_2 is the dominant GHG resulting from operation of natural gas and coal-fired plants, full fuel-cycle GHG emissions include N_2O and CH_4, primarily from upstream processes related to the production and transport of the fuel source.

- Three scenarios represent high, medium, and low penetration of PHEVs in the 2010 to 2050 timeframe.

From these two sets of scenarios emerge nine different outcomes spanning the potential long-term GHG emissions impacts of PHEVs, as shown in the following table.

Table 4-1. Annual Greenhouse Gas Emissions Reductions from PHEVs in the Year 2050

2050 Annual GHG Reduction (million metric tons)		Electric Sector CO_2 Intensity		
		High	Medium	Low
PHEV Fleet Penetration	Low	163	177	193
	Medium	394	468	478
	High	474	517	612

The EPRI-NRDC researchers drew the following conclusions from the modeling exercises:

- Annual and cumulative GHG emissions are reduced significantly across each of the nine scenario combinations.

- Annual GHG emissions reductions were significant in every scenario combination of the study, reaching a maximum reduction of 612 million metric tons in 2050 (High PHEV fleet penetration, Low electric sector CO_2 intensity case).

- Cumulative GHG emissions reductions from 2010 to 2050 can range from 3.4 to 10.3 billion metric tons.

- Each region of the country will yield reductions in GHG emissions.

SMART CHARGING

There has obviously been a major shift in the automotive industry's attitude towards plug-in hybrid and electric vehicles during the last two years. As the plug-in electric vehicle (PEV) market grows, the need for utility/end-user interaction grows with it. A study in Nashville, Tennessee, discovered that the customers most likely to purchase electric vehicles lived in some of the neighborhoods least capable of supporting the additional demand. Adding an electric vehicle to a household has been compared to adding an entire house. During the charge cycle, the PEV can draw more current than all the other appliances in the home combined. The utilities' ability to communicate, monitor, and control PEV charging must become a reality to minimize the negative impacts on the distribution system.

Smart Charging Plug-in Electric Vehicles
The first plug-in electric vehicles have been announced and will be in production starting with late 2010. These vehicles can be characterized by a large (10 to 15 kWh) on-board battery that is expected to be charged overnight and at home most frequently. At the same time, with the cordset and on-board charger, at least some of the owners may make them plug in their PEVs to any available standard 120-volt or 240-volt outlet anywhere, with the intent to use least amount of liquid fuels and to be environmentally conscious. There are certain times of day when

it is more preferable for PEVs to be charging from the grid (and certain others, that are not as preferred). Therefore, the vehicle owners and utility companies can benefit the most from having a PEV if the vehicle was compliant with one of the energy efficiency or demand response programs. Even more so if the PEV can actually verify its participation through some means. This would necessitate the ability of the utility system to somehow control, when required, the timing and amount of energy that the PEV draws from the grid, and then perhaps reward the owner for doing so through discounted rates, for example.

Several federal acts promote or mandate consideration of PEVs. The Federal Energy Policy Act (EPAct) from 2005 promoted efficient energy utilization. The Energy Independence and Security Act (EISA) of 2007 that mandates the grid to be secure and intelligent, several nationwide initiatives to add intelligence to the power grid management and control. This has resulted in several initiatives which are underway, including the IntelliGrid project from EPRI, Smart Grid projects at various utilities and Department of Energy, as well as Automated Metering Infrastructure installations at a wide range of utilities across North America. All of these have a common goal of embedding intelligent power and energy management systems into the power delivery infrastructure to help monitor, manage and control the power flow and delivery all the way to the end-use devices, with the objective to achieve the best throughput at the least operating costs.

These "smart grid" systems are characterized by three elements in general—(1) a sensor system that determines the current state of the grid, (2) an intelligent power management system that determines and communicates the desired state of the system to the smart loads, and (3) a network of smart loads that receive these communications, act on the command, and report their current state to the smart grid. Therefore, to enable this vision of "smart grid," the related initiatives are focused on all three aspects, of which the first two (sensors and intelligent power management) reside within the grid, while the third, the smart loads, are the end-use devices such as programmable/controllable thermostats (PCTs), home energy management systems, smart refrigerators, washer/dryers, and plug-in electric vehicles.

The "smart" part of the smart loads is both their ability to communicate as well as process the control signals received from the "smart grid," which include both the communications functionality and the control algorithms that allow the battery charging to be controlled and feedback signals be computed for communicating with the smart grid. This functionality also applies to plug-in electric vehicles, for their owners to avail themselves of the maximum benefits of using electricity as a transportation fuel. Since PEVs are one such class of smart loads, they inherit the requirement to be able to communicate and control power flow from the grid, as commanded by the smart grid. Collectively, these technologies enable what is termed as "smart charging" functionality.

Current State of Smart Charging Technology

While until as recently as 2007, the concept of the PEV's control and communication system connected to the grid was foreign to the automobile manufacturers, this situation is rapidly evolving and in the span of last few months, a whole host of initiatives have been established to quickly address this gap in automobile's capabilities to communicate with the grid and control its own energy draw from the grid as needed, as well as provide acknowledgement of having done so.

Smart Charging—Drivers

A confluence of several factors is driving the need for smart grid wanting to "talk" to smart charging plug-in electric vehicles. These are:

- Concerns over global climate change and a need to optimize the carbon footprint, resulting in a need for improving the load factor of existing capacity.
- A national desire to achieve energy independence and energy security through relying on home-grown sources of energy, preferably clean.
- A consumer sentiment wanting to pursue cheaper and cleaner sources of transportation energy, i.e., electricity, in light of price of crude oil almost permanently stationed above \$100/bbl, and retail gasoline prices approaching \$4/gallon.
- A maturation of electric drive and energy storage technologies in terms of their reliability, robustness, size, weight and cost economics, owing to the widespread adoption of hybrid electric vehicle technologies.
- A desire for automobile manufacturers to both comply with applicable clean-air regulations as well as one-up each-other, as a means to bolster their environmental and technology credentials and image.

The net result is that two of the largest industries in the North America—automotive and utilities—are simultaneously undergoing transformational changes that occur may be once in several decades.

Auto Industry Initiatives Around Plug-in Electric Vehicles

The automotive industry has undergone a massive shift in its attitude towards hybrids, and more specifically, plug-in hybrid and electric vehicles in just the last two years, since the announcement of Chevrolet Volt and Saturn VUE plug-in electric vehicles from General Motors Corporation, and the Ford Escape Plug-in Hybrid Electric Vehicle (PHEV) demonstration program in conjunction with Southern California Edison. Europeans and some Japanese automotive OEMs, not to be outdone by these developments, have announced their own production and demonstration programs for PHEVs and PEVs.

While all of these OEMs are choosing to have their products going through real-world market and usability tests, their reasons for doing so are varied and can be categorized as follows:

- Compliance with Zero Emission Vehicles Mandate from the state of California and 14 states on the east coast.

- Competitive response for technology image and environmental stewardship.

- Pure public relations exercise.

- Making the technology ready through real-world experimentation.

- Establishing technological leadership to change public perception about the company's products and capabilities.

Regardless of the strategic objectives of the OEMs, as long as their vehicles require energy to be drawn from the grid, the utility industry has an interest in ensuring that the charging of these vehicles is done in a benign manner at a minimum, and ideally in a manner beneficial to the grid. This is why there is a tremendous active interest from utility industry and EPRI in developing the technologies pertaining to the smart charging together, for joint deployment on both sides of the plug.

TRUCK STOP ELECTRIFICATION

Not all opportunities for CO_2 reduction will come from new plug-in electric vehicles. One exciting opportunity regarding existing vehicles is truck stop electrification.

About 1.3 million of the nation's 2.5 million heavy-duty trucks are equipped with sleeper berths and operate on U.S. highways and transportation corridors at any one time (EPRI 1008777). Drivers of long-haul trucks are required by the U.S. Department of Transportation to comply with truck driver hours-of-service regulations. These regulations fundamentally allow an 11 hour driving period after a 10-hour rest period. During this rest time, drivers idle their truck to maintain sleeper cab comforts. Each long-haul trick idles an average of 1,830 hours per year. Shorter idling periods occur at warehouses, truck fleet terminals and distribution centers, while longer overnight idling periods occur at designated truck stops.

This idling creates significant pollution and wastes fuel. As a fleet, sleeper truck idling consumes more than 915 million gallons of fuel annually (EPRI 1008777). In addition, truck idling emits harmful air pollutants, totaling a combined 272,000 tons per year of oxide of nitrogen (NO_x), particulate matter (PM), hydrocarbons (HC) and carbon monoxide (CO), with an addition 9.2 million tons per year of carbon dioxide (CO_2), a greenhouse gas.

Each year an idling sleeper truck:

* Consumes an average of 1,830 gallons of diesel fuel per truck.

* Emits 18.4 tons of CO_2 per truck.

* Emits 0.33 tons of NO_x per truck.

* Emits 11.3 pounds of diesel particulate matter, which adds up to nearly 2,800 tons per fleet.

* Causes significant noise disturbances for local residents.

Additionally, sleeper truck idling necessitates more frequent maintenance—shorter oil change and engine overhaul intervals. Truck stop electrification (TSE) provides facilities which allow long-haul truckers to "plug in" their vehicles at a truck stop. These plug-in facilities allow the operation of air conditioning, heating and appliances without engine idling.

Per Truck Equivalency

Replacing one sleeper truck's idle hours per year with electric power reduces the same amount of NO_x emissions as removing 360 passenger cars from the road each year, and saves an average of 1,830 gallons of diesel fuel, annually. In addition, the CO_2 reductions per year are the equivalent of removing three cars from the road or planting over 1,500 full-size trees (EPRI 1008777).

Per Space Equivalency. Electrifying one truck stop parking space that is used 16 hours per day (5,840 hours per year, assumes multiple shifts per day at each space) can provide NO_x reductions equal to removing nearly 1,150 new passenger cars per year, and can cut annual diesel fuel consumption by more than 5,840 gallons. Electrifying one space can also reduce CO_2 emissions equal to removing nearly 10 cars from the road or planting nearly 5,000 full-size trees annually.

TRUCK DRIVER BENEFITS FROM TSE

TSE allows truck drivers to turn off their engine and still maintain cab comforts. Drivers benefit in the following ways:

- No engine vibration or noise.

- No exposure to harmful emissions.

- No need for an auxiliary power unit (APU) that cycles on and off during the rest period.

- Extended oil change intervals and engine life.

- Significant annual fuel cost savings ($3,220 on average per year).

- The ability to use appliances without an adverse affect on truck charging and battery systems.

TWO MAJOR TYPES OF TRUCK STOP ELECTRIFICATION

Off-board TSE systems have extensive infrastructure, including HVAC systems. Any of today's sleeper cabs, including more modern sleeper cabs that are shore-power-equipped with plugs and on-board HVAC, can use off-board TSE systems. The off-board HVAC is located in a structure above the truck parking spaces. A hose from the HVAC system and a module containing AC power outlets and Internet connection are connected to the truck window, while a computer touch screen provides control, enables payment, and offers premium entertainment options. Stand-alone systems are owned and maintained by private companies that charge an hourly usage fee.

Shore-power TSE systems have on-board HVAC systems (and other amenities) and plugs on the truck so the truck can use AC power from any 120 V outlet. However, many places that trucks idle do not have convenient outlets. Infrastructure for this type of TSE is much less extensive than with off-board TSE. The main challenge for shore-power TSE systems is the modification of sleeper cabs or facilitating the ability for new sleeper cabs to be appropriately equipped. Current industry trends suggest that trucks will, in the future, be equipped with diesel auxiliary power units and wired to accept shore power. This preserves the ability to run appliances when shore power is not available, while providing the opportunity to take advantage of lower cost and lower maintenance shore-power when available.

SOCIETAL BENEFITS FROM TSE

There are many important societal benefits associated with reduction of extended (i.e., overnight) truck idling including:

- Toxic air pollutant reductions including formaldehyde and diesel PM.

- Health and safety benefits of well-rested truck drivers.

- CO_2 emission reductions, important in the effort to mitigate global climate change.

- Fuel consumption reductions resulting in a decrease in foreign oil import dependence.

- Cost savings (i.e., fuel savings, decreased maintenance costs, and longer engine life) to truck owners that translate to lower cost of goods.

- Reduced noise at truck stops, distribution centers and other areas.

- Responding to environmental justice concerns since truck stops are often located near low-income and minority populations.

EMISSION REDUCTION BENEFITS OF TSE

Table 4-2 summarizes the range of emission and fuel reduction benefits achievable from idle reduction that uses TSE, based on low and high case utilization rates. The low case is conservative and is based on a usage of 5 hours per day, 365 days per year. The high case is based on an assumed 16 hours per day of idling, previously noted. Of course, as the trucking industry's interest in TSE grows, so will its utilization.

Table 4-2. Range of Benefits From TSE

Description	Los Case: 1,825 Hr/Yr (One Truck per Space)	High Case: 5,840 Hr/Yr (Multiple Trucks per Space)
PM (Pounds/Yr) Pre-2007	11.14	35.66
PM (Pounds/Yr) 2007 & Later	1.13	3.61
CO$_2$ (Ton/Yr)	18.4	58.84
NOx + HC (Ton/Yr	0.36	1.14
Diesel Gallons per Year	1,825	5,840

CONCLUSIONS

The transportation sector overall uses approximately one-third of all energy—mostly through fossil fuels. Yet less than 1% of electricity in the U.S. is used to power transportation systems. And most of them are electrified rail applications. Clearly, on-road electric vehicles hold great promise for reducing overall energy use and reducing CO$_2$ emissions.

References
"Environmental Assessment of Plug-In Hybrid Electric Vehicles, Volume I: Nationwide Greenhouse Gas Assessment," Electric Power Research Institute (EPRI) and the National Resources Defense Council (NRDC), EPRI 1015325, July 2007.
"Truck Stop Electrification: A Cost-Effective Solution to Reducing Truck Idling," EPRI, Palo Alto, CA: 2004. Technical Brief 1008777.
"Smart Charging Development for Plug-In Hybrid and Electric Vehicles—Preliminary Use-Case Development for SAE Recommended Practice J2836," Electric Power Research Institute, Palo Alto, CA: December 2008. Technical Update 1015886.
"Owning an Electric Car," Michael Boxwell, Greenstream Publishing, 2010.
"John's Stuff—Toyota Prius History," http://john1701a.com.
"An Early Road Warrior—Electric Vehicles in the Early Years of the Automobile," C. Sulzberger, IEEE Power and Energy Society, www.ieee.org, 2004.
"Nissan Leaf," http://en.wikipedia.org/wiki, Nissan-Leaf, 2010.

Chapter 5

Electrifying Off-Road Motive Power

Off-road motive power is used in a variety of non-road or off-road applications. These applications are comprised of vehicles and equipment which encompass a broad range of technologies. The applications of these technologies may fall into the following segments: construction and mining; industrial; lawn and garden; farm; commercial; logging; airport service; railway; and recreational. A convenient grouping of the most important electrification applications among these is as follows:

- **Industrial Equipment**: forklifts, sweepers/scrubbers, varnishers, and other material-handling equipment.

- **Airport Support Equipment and Vehicles**: ground support equipment used in airport operations including equipment for maintaining and fueling aircraft, transporting and loading cargo, transporting passengers, handling baggage, servicing lavatories, and serving food.

- **Seaport**: land-side equipment including canes, forklifts and electric vehicles, cold ironing on ships, and electric locomotives.

- **Lawn and Garden Equipment**: lawn mowers, weed trimmers, brush cutters, leaf blowers/vacuums, riding mowers, chainsaws, tillers, shredders, lawn and garden tractors, snow blowers, chippers/stump grinders, and commercial turf equipment.

- **Railroads**: electric locomotives, light rail vehicles, and magnetic levitation systems.

- **Mining**: conveyors, shuttle cars, draglines, electric shovels, and electric locomotives.

Emissions from fossil-fueled versions of these vehicles and equipment include oxides of nitrogen (NO$_x$), reactive organic gases (ROG), carbon monoxide (CO), particulate matter (PM), greenhouse gases (CHG), and air toxics. In addition, exhaust emissions from these vehicles are harmful to human health and damaging to the environment (EPRI 1002244). According to EPA, the nearly six million non-road fossil-fueled engines in the U.S. contribute 12% of the total NO$_x$ emissions and 44% of the total PM emissions. Many non-road vehicles are significantly dirtier than on-road vehicles. Federal, state, and local agencies are aggressively trying to mitigate the emissions and encourage the implementation of low-polluting technologies. Aspects of this work include the promulgation of standards that limit the emissions allowed by mobile sources. Figure 5-1 illustrates CO$_2$ emissions from off-road sources.

This chapter will describe some of the most important opportunities for electrifying off-road, fossil-fueled motive power.

Eladio Knipping at EPRI has compiled an excellent list of equipment and potential electrification technologies which could be considered as off-road applications of beneficial electrification. Table 5-1 summarizes his findings.

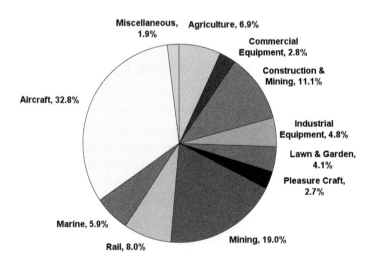

Figure 5-1. U.S. CO$_2$ Emissions From Off-Road Sources

Table 5-1. Off-Road Equipment Types Included in the Electrification Screening Analysis (Source: Knipping)

Equipment Type		Technology/Technologies
Industrial Equipment		
1.1	Forklifts	battery*, hybrid
1.2	Sweepers/scrubbers	battery*
1.3	Transportation refrigeration units	plug-in
Construction/Landfill/Mining		
2.1	Excavators	hybrid
2.2	Loaders/dozers	hybrid
2.3	Off-highway trucks	hybrid, battery*
2.4	Signal boards	battery*, plug-in
Locomotives		
3.1	Switching	battery*
3.2	High- speed rail/magnetic levitation	Maglev electric conveyor
Cargo Handling Equipment		
4.1	Cranes	plug-in
4.2	Yard hostlers	battery*
Marine		
5.1	Shore-side power	plug-in
5.2	Tugboats	hybrid
5.3	Dredging	plug-in
5.4	Ferry and excursion vessels	hybrid
Lawn and Garden Equipment		
6.1	Mowers	battery*, plug-in
6.2	Leaf blowers	battery*, plug-in
6.3	Trimmers/edgers	battery*, plug-in
6.4	Chippers/shredders	plug-in
6.5	Snow blowers	plug-in
6.6	Chain saws	battery*, plug-in
6.7	Lawn and garden tractors and riding mowers	battery*
6.8	Commercial turf equipment	battery*
Recreational Equipment		
7.1	ATVs	battery*
7.2	Motorcycles	battery*
7.3	Special vehicle carts	battery*
7.4	Golf carts	battery*

(Continued)

Table 5-1 (*Cont'd*). Off-Road Equipment Types Included in the Electrification Screening Analysis (Source: Knipping)

Equipment Type	Technology/Technologies
7.5 Snowmobiles	battery*
7.6 Personal watercraft	battery*
Airport Ground Support Equipment	
8.1 Auxiliary power units	plug-in
8.2 Baggage tug	battery*
8.3 Belt loader	battery*
8.4 Cargo loader	battery*
8.5 Pushback	battery*
Oil and Gas Field Equipment	
9.1 Pumpjacks	plug-in
9.2 Compressors	plug-in
9.3 Pneumatic pumps	plug-in
9.4 Pneumatic devices	plug-in
Agriculture Equipment	
10.1 Pumps	plug-in

FORKLIFTS

Battery-powered forklifts have the largest market share of any non-road electric vehicle technology. The drivers for this growth include health and safety regulations, productivity advantages, and favorable life-cycle costs. There is room for expanded use in less traditional industries such as construction, lumber, cargo handling and metal work.

Forklifts are one of the major types of non-road vehicles. Forklifts are defined as mobile vehicles powered by electric motors or internal combustion engines and used to carry, push, pull, lift, stack, or tier materials. The Industrial Truck Association (ITA) has defined seven classes of forklifts. These classes are characterized by the type of engine, work environment (indoors, outdoors, narrow aisle, smooth or rough surfaces), operator positions (sit down or standing), and equipment characteristics (type of tire, maximum grade, etc.). Several classes are further divided by operating characteristics. Table 5-2 lists the forklift classifications as well as available information from various sources on typical fuels, characteristics, horsepower, and lift capacity (EPRI 1002244).

Electric forklifts dominate both in the lower lift capacity portion of the market and in indoor applications. They have been less accepted for outdoor and heavy-duty applications. The perceived limitations of electric forklifts have been:

- Battery range and performance

- Water-proofing for outdoor applications

- Speed

- Gradeability (ability to handle grades)

Table 5-2. Forklift Classes

Class	Lift Code	Engine Type	Type/Use	Typical Lift Capability	Typical hp
1	1	Electric	Counterbalanced rider, stand up		
1	4		Three-wheel, sit down		
1	5		Counterbalanced rider, sit down	3,000-6,000 lbs.	• 50
1	6		Counterbalanced rider, sit down	3,000-6,000 lbs.	• 50
2			Harrow aisle truck	3,000-6,000 lbs.	• 50
3			Hand or hand/rider truck	3,000-6,000 lbs.	• 50
4		ICE* – gasoline, CNG,** propane, diesel	Rider, sit down, generally suitable for indoor use on hard surface	3,000-16,000 lbs.	50-120
5			Rider, sit down, typically used outdoors, on rough surfaces or steep inclines		50-120
6		ICE* – gasoline, CNG,** propane, diesel, electric	Ride-on unit with the ability to two at least 1,000 pounds; designed to two cargo rather than lift it (e.g., an airport tug)		>75
7		ICE* (primarily diesel)	Rough Terrain forklift truck for outdoor use; almost exclusively powered by diesel engines	6,000-40,000 lbs.	> 750

* ICE – internal combustion engine; also referred to as internal combustion (IC)
** CNG – compressed natural gas

- Performance at higher lift capacities

As a result of these real or perceived limitations of electric forklifts, multiple shift forklift users lean toward forklifts with internal combustion engines (ICE). However, recent technological developments are overcoming these barriers to the expansion of electric forklifts into non-traditional markets. Recent performance improvements include:

- Increased travel speeds and acceleration rates competitive with ICE forklifts
- Increased ramp speed and gradeability
- Pneumatic tire designs for outdoor operation
- Rust-proof designs for operation in temperature-controlled and marine environments
- Sealed spark-proof and explosion-proof designs for operation in combustible environments
- Increased battery performance
- Sealed maintenance-free batteries
- Regenerative braking

These improvements, plus other advances such as more comfortable seating and more joystick-type gears, provide an opportunity for electric forklifts to penetrate new markets.

Other Advantages to Electric Forklifts

The health and safety benefits of electric forklifts are significant. Electric forklift technologies are available today and are widely accepted. Various worker health and safety rules (Occupational Safety and Health Administration—OSHA) now require that forklifts in certain enclosed environments be electric. There are substantial health and safety benefits to electric forklifts for workers in less restrictive environments. The health effects of prolonged exposure to diesel emissions even in outdoor environments are widely known to be unhealthy. Studies of railroad, seaport dock, trucking and bus garage workers exposed to high levels of diesel exhaust over many years demonstrated a 20 to 50% increase in risk of lung cancer or premature death (Union of Concerned

Scientists web site, www.ucsusa.org).

Current emissions models account for tailpipe emissions only and do not take into consideration the lower life-cycle costs and longer emissions benefits of electric forklifts. Even with the lack of full fuel-cycle or life-cycle data, practitioners typically find electric forklifts to be the cleanest technology.

Electric Outdoor Lift Trucks

Electric lift trucks are more commonly used indoors and have not significantly penetrated the outdoor market because customers have perceived them to be unsuitable for outdoor use. Increasingly, manufacturers are building trucks with powerful new alternating current (AC) drive technologies coupled with features tailored to outdoor use. Many major manufacturers offer electric lift trucks suitable for outdoor use (EPRI 1020515).

Lift trucks equipped for outdoor work differ from indoor lift trucks. Outdoor lift trucks include pneumatic tires that improve ride and handling on rough and uneven surfaces, enclosed motors and electronic systems to ensure safe operation in inclement weather and heavy dust, and optional climate control of the operator compartment. Indoor trucks use solid or cushion tires, do not require weatherproofing, and may require less lifting capacity. Powerful AC drives and higher voltage systems deliver the speed, handling, and performance needed for demanding outdoor use.

From 2007 through 2009, EPRI, with utility members Southern Company (Alabama Power and Georgia Power) and New York Power Authority, partnered with lift truck dealers to demonstrate electric out-

Table 5-3. demonstrates the life-cycle costs for electric material-handling vehicles as compared to fossil-fueled vehicles.

	Electric	Propane (LPG)	Gasoline	Diesel
Life Expectancy	9 years	6.5 years	6.5 years	7 years
Initial Cost (mfg. suggested list price)	$29,739	$21,200	$20,107	$22,263
Annual Fuel Costs	$585	$2,917	$3,463	$2,115
Annual Maintenance Cost	$1,655	$2,800	$2,800	$2,800
Total Life Cost	$49,512	$58,364	$60,814	$56,671
Annual Average Cost	**$5,570**	**$8,979**	**$9,357**	**$8,096**

door lift trucks in more than 45 locations. The objectives of the demonstration were to penetrate the 6,000-pound or less IC outdoor lift truck market, demonstrating to the end user that electric outdoor trucks perform equally as well, or in some cases better than, IC trucks, and to demonstrate that electric outdoor lift trucks not only operate as efficiently as their IC counterparts, but also greatly reduce fuel costs and emissions.

The biggest barriers to electric outdoor lift truck use are misperceptions and a lack of information. Users perceive that electric trucks are underpowered, that batteries cannot last a full shift, and that electric trucks are unsafe in wet weather. Most users have no experience with electric outdoor trucks, they do not hear about the trucks' benefits from dealers, and they have no economic payback information. As a result, they do not demand electrics from their dealers.

For their part, dealers have not recommended electrics for similar reasons: they have little or no experience with the equipment (or their past experience was unsatisfactory), they believe it has lower performance, and customers do not demand it. Additionally, because electric outdoor trucks cost more up front, dealers are reluctant to carry them

Figure 5-2. An Electric Outdoor Lift Truck Drives Up a Ramp With Ease at Springer Equipment Company in Birmingham, AL. (Source: Alabama Power Company)

in inventory, further reducing their and their customers' exposure to electrics.

AIRPORT ELECTRIFICATION

There are six categories of energy use at airports which produce harmful emissions. These include the following:

- **Aircraft Activity**: Emissions from commercial aircraft general aviation, and air taxi aircraft are the predominant contributors of airport nitrous oxides (NO_x) and sulfur oxides (SO_x) and are significant sources of hydrocarbons (HC). Aircraft activity also generates substantial quantities of carbon monoxide (CO).

- **Auxiliary Power Units (APUs)**: APUs that generate electricity and compressed air on board the aircraft to operate aircraft instrumentation, lights, ventilation, and other equipment during main aircraft engine shutdown, generate small quantities of airport CO, HC, and NO_x, and negligible quantities of SO_x and particulate matter (PM) while operating at the airport.

- **Ground Support Equipment (GSE)**: Equipment servicing commercial aircraft while passengers board and deplane are the predominant contributors of CO. Such equipment also contributes to HC, NO_x, SO_x, and PM airport emissions.

- **Airport Support Vehicles (ASVs)**: Air- and land-side ground vehicles that support the day-to-day operations of airports are the major generators of airport PM emissions. These include emergency response vehicles, shuttle buses, and airport authority sedans, vans and pickups. Depending upon the airport, ASVs can also be significant contributors of CO, HC, NO_x, and SO_x emissions.

- **Ground Access Vehicles (GAVs)**: All on-road vehicles traveling to and from, as well as within, the airport can be significant contributors of CO, HC, and PM emissions. They also generate small quantities of NO_x and SO_x.

- **Stationary Sources**: On-site combustion sources such as combined heat and power systems, boilers and emergency generators, as well as non-combustion sources such as fuel storage and handling operations, comprise a small percentage of total airport emissions.

There are a number of electrification strategies which can be employed at airports to achieve emissions reductions. Airport emissions reductions from electrification are technically limited only by the availability of direct or indirect replacement electric vehicles and equipment. Economically feasible emissions reductions are typically limited to electric vehicles and equipment that were both technically viable and have life-cycle costs equal to or less than those of their internal combustion counterparts.

The following electrification strategies are feasible:

- **Aircraft Taxiing Using Electric Tugs**: The use of electric tugs to tow incoming aircraft into gate areas during travel times could potentially reduce total airport CO, HC, and NO_x by about 1%.

- Use of **Electric Ground Power and Pre-Conditioned Air Supply** in Place of Auxiliary Power Units: Limiting the time in which APUs operate at gates could potentially reduce total airport CO, NO_x and HC emissions. This strategy could result in a slight increase of SO_x emissions.

- **Electrification of Ground Support Equipment (GSE)**: The replacement of internal combustion GSE (belt loaders, GPUs, pushback tractors, and tugs) with currently available electric models could potentially provide the greatest emissions reduction potential for CO, HC, and NO_x. This strategy could either result in a modest increase in total SO_x emissions or an overall reduction, depending on the electric fuel mix used in the local grid supplying the airport.

- **Electrification of Airport Support Vehicles**: The replacement of internal combustion airport support vehicles such as carts, forklifts, portable stairs, work platforms, and shuttle buses could potentially reduce CO, HC, NO_x, and PM. This strategy could either result in a modest increase in total SO_x emissions or an overall reduction, depending on the electric fuel mix used in the local grid supplying the airport.

Other airport electrification strategies involving stationary sources are also possible. Such strategies could include electrification of terminal and auxiliary building systems and equipment, as well as the electrification of ground access vehicles. The likelihood of electrifying passenger vehicles may be considered in long-term plans. Ground access vehicle emissions reductions could also be accomplished through the use of electric mass transit systems that would reduce the number of motor vehicles traveling to and from airports.

According to a study by EPRI (EPRI 111000), significant reductions in emissions could be realized if all four of the emissions reduction strategies mentioned earlier were implemented. The total economically viable emissions reduction ranges of the airport were calculated as follows:

- CO (48 to 56%)
- HC (18 to 24%)
- NO_x (7 to 17%)
- PM (16 to 44%)
- SO_x (024 to 14%)

In the same study, EPRI found changes in SO_x emissions ranged from an increase of 22% (an increase in emissions, albeit they occur at the power plant and not the airport) to a reduction of 14%. EPRI also found that an increase in SO_x emissions can occur because of high oil and coal utility fuel mixes, which emit more SO_x per unit of energy than vehicles or equipment powered by gasoline (which has a very low sulfur content).

Electric Container Pallet Loader

Electrification of airport ground support equipment (GSE) is one way to reduce operating costs and emissions at airports. An electric-drive option exists for many GSE models, and more options are being introduced every year. However, replacing IC units with new electric-drive equivalents may not make economic sense when existing units have not yet reached end of life, unless environmental constraints are paramount. One solution is to retrofit existing equipment to electric drive.

Benefits include significant savings and reduced emissions: Retrofitting existing equipment costs about one-fifth the cost of replacing old equipment, and electrification results in operating cost savings of approximately 80%. Additionally, electric-drive equipment aids compliance with emissions regulations (EPRI 1020482).

For example, a container pallet loader conversion kit has been developed for a diesel-powered aircraft cargo loader. The cargo loader is used in airports worldwide to move luggage containers and other palletized cargo between ground level and cargo bins on wide-body aircraft. The duty cycle of these devices is typically short duration under heavy load, limited distance traveled, and a significant amount of downtime between servicing aircraft.

Combined Heat and Power

Airports offer a great opportunity for on-site generation of heat and power. As they operate around the clock, they require optimal power quality and reliability, their annual electricity consumption is high (~10,500 MWh), their average electric load during operating hours is considerable (~1.2 MW), and the ratio of their thermal to electric load is high (~5.6:1). Energy costs are significant at airports, accounting for 10 to 15% of operating costs. A modular, cost-effective system for on-site generation of electricity and heat could greatly benefit airports by supporting general operations, as well as by facilitating the conversion of land-based vehicles used at airports to electric-driven vehicles. Such a system could also enable the on-site production of hydrogen for use as a fuel alternative in airplanes. Virgin Atlantic/Branson is actively pursuing alternative jet fuels. Currently, hydrogen is being investigated for hypersonic airplanes. The market for alternative fuels could be significant. For reference, the energy use by certified carriers in the U.S. was 560 million MWh in 2003 including all fuels (no hydrogen at this time). Because a typical airport does not require 25 MW for operation, the best scenario for justifying this size of generation system would be to use the excess electricity for generating hydrogen. Depending on the on-site hydrogen demand, any remaining hydrogen could be sold and distributed for other off-site applications.

SEAPORT ELECTRIFICATION

Like many industries across the U.S., seaports have begun investigating ways to reduce air emissions associated with their operations. As ports consider their emission reduction options, they must first consider the control strategy opportunities available to them to reduce emissions. Equipment electrification is one strategy which could reduce air emissions associated with seaports (EPRI 1010577).

The impetus for seaports to reduce emissions comes from a variety of different drivers including:

- An effort to be good neighbors and good citizens.
- A response to plans to decrease state-wide emissions in compliance with state implementation plans and other programs.
- To comply with federal non-road vehicle standards.
- The ability to expand operations.

There are several different control strategies available to reduce emissions depending on the area of port operations and the pollutants involved. In considering control strategies for land-side equipment, for example, options include:

- Equipment retrofit devices
- Repowering with cleaner engines
- Replacing equipment
- Electrification

Among the variety of options available to ports, electrification can achieve not only substantial emissions reductions but can also be operationally sound.

There are several electrification categories within a seaport that can potentially utilize electric power to reduce emissions; these include:

- Electric land-side equipment including cranes and forklifts
- Electric and hybrid electric on-road vehicles operating within ports
- Truck stop electrification
- Cold ironing on ships
- Electric locomotives

There are electric options for cargo handling equipment, such as cranes and forklifts, used to load goods on and off of ships and around terminal yards. Vehicles that operate at the port may also be electric and/or hybrid electric. Trucks used for cargo transport may be hybrid electric-diesels. Ships and terminals can "cold iron," or plug into shore power, while in port instead of using diesel power generators to supply power for the ship. Switching locomotives that move cargo in and around ports can also be modified to be electrified, although the most popular alternative to date is the hybrid battery locomotive having a small diesel auxiliary engine (EPRI 1010577).

Seaport land-side equipment is largely run on diesel fuel. However, electric alternatives are available for most land-side applications.

Electricity Options for Land-Side Equipment

In general, there are several ways to convert land-side internal combustion industrial equipment to electric use.

- Direct hookup to the power grid.
- Battery power—Internal combustion engines are replaced with electric motors which are powered by batteries. Forklifts are ideal for this technology.
- Hybrid electric vehicles—Hybrids can use small diesel engines used for boost power and to recharge on-board batteries. Examples include switch locomotives and battery-powered industrial equipment having micro-turbines to continuously recharge batteries.

The first two options, direct electric and battery power, are particularly attractive because the non-road emissions, especially for oxides of nitrogen and particulate matter, are eliminated. Despite the fact that the power used to supply this equipment is attributed to the power source, which could be fueled by coal, nuclear, hydro, wind, or natural gas, and as such have associated emissions. but because the electric equipment is far more effective at the point of end use and increasing amounts of electric power production is from low-carbon sources, the net reductions in emissions is substantial.

Hybrid systems usually involve some use of an auxiliary diesel engine that has some associated emissions. Nonetheless, there can still be significant savings.

Some cargo handling machinery is described as "diesel-electric." This means that a diesel engine is used to drive a generator, which in turns creates power for electric-drive motors. Machinery of this kind is still considered as a diesel engine with its associated emissions.

There are several types of electric equipment that are commercially available, either as a retrofit of an existing piece of equipment or from a new original equipment manufacturer (OEM) including:

- Wharf gantry cranes
- Railcar movers
- Forklifts
- Man-lift devices
- Railcar/container loading gantry crane
- Light-duty vehicles
- Lighting powered by the grid (much of the lighting used for night work at seaports is diesel powered)
- Refrigerated containers (also know as reefers)

In addition to electric equipment that is commercially available, there are two new equipment trends with regard to container operations at seaports: the automation of gantry cranes and the automation of terminal tractors or yard trucks.

Yard Tractors

Diesel-powered tractors are trucks similar to the cab on a tractor-trailer that pull containers and cargo are used extensively in ports and warehouses, where it is necessary to shuttle cargo trailers from point to point within the confines of a specific facility, terminal or yard. Often called yard tractors, yard hostlers or terminal tractors, this equipment is of a specific design with a single driver compartment and a fifth wheel. These terminal tractors are unique to the cargo industry (EPRI 1017678). Seaports are under pressure to reduce emissions. Ports are beginning to take great steps in reducing emissions from various aspects of their operations. Many emission-reduction efforts focus on cargo-handling equipment such as the terminal tractor due to this sector's share in port emissions and concerns over oxides of nitrogen (NO_x) and particulate matter (PM) emissions. Electric technology can play an important role in this target sector by reducing equipment emissions.

At ports, where cargo-handling equipment is used throughout container terminal operations, terminal tractors can comprise a large portion of all container terminal cargo-handling equipment. At large container ports like the Port of Long Beach, California, the number of terminal tractors at the container terminal alone is approximately 750 vehicles (EPRI 1017678). In this environment, terminal tractors are used during the entire operational shift, but typically idle a large portion of their shift while waiting to pick up a load and shuttle it to another area of the port. Such idling results in increased engine emissions and unnecessary fuel consumption.

The development of plug-in hybrid tractors are under development which address energy and environmental concerns by allowing the main drive engine to be shut off during idling, effectively mirroring the hybrid vehicles on the road today.

Electric Ship-to-Shore Cranes

A ship-to-shore (STS) crane is a large dockside crane that moves containers to and from ships at container ports. STS cranes have a lifting device attached to them called a spreader that picks up and moves containers. Container cranes are classified by their lifting capacity and the size of the containers on the ships they can load and unload. A modern container crane capable of lifting two 20-foot-long containers at a time will typically have a rated lifting capacity of 65 tons.

These cranes are commonly available in either diesel or electric mode, stay in a well-defined area of the dock, and typically work and idle constantly while a container ship is at port. An average crane diesel engine with a 1,500-horsepower (HP) engine running 3,800 hours per year uses approximately 69,000 gallons of diesel fuel; the calculated electric equivalent uses 890,000 kilowatt-hours (kWh) of electricity (EPRI 1020510).

Although in the past, STS cranes were largely powered by diesel engines, they are now often powered by electric motors. These typically use electric power from the dock with an electrical service requirement ranging from 4,160 to 13,800 volts.

The motivation for this move toward electric STS cranes has been three-fold: (1) the recognition of economic, energy, and operational efficiencies of electric compared to diesel; (2) rising fuel prices; and (3) emission-reduction pressures at ports.

STS cranes are workhorses for container ports, sometimes being used around the clock; therefore, any efficiencies in the operation of the crane can be associated with benefits. Although available at a comparable price to its diesel counterparts, the electric STS crane can have:

- Greater energy efficiency resulting in reduced operating costs
- Enhanced reliability with respect to diesel units
- Lower maintenance costs than diesel units
- Longer equipment life
- Emission reductions

- Enhanced safety

Electrification Associated with Ships—Cold Ironing

Ships docked at port usually shut off their propulsion engines but still run their auxiliary diesel generators to power ventilation, refrigeration, lights, and pumps. The emissions associated with this activity, commonly called "hotelling," can be substantial depending on the size of the ship, quantity of ships in port, and the duration they are there.

In order to reduce the emissions associated with hotelling, ships at berth can "cold iron," defined as the use of shore power by the ship instead of its own diesel generators for at-dock needs. Substantial emission reductions can be achieved through cold ironing. However, it is comparatively expensive and requires substantial infrastructure upgrades, both on shore and aboard the vessels that cold iron.

There are currently no international requirements that mandate cold ironing, and there are currently very few cargo ships that cold iron at U.S. ports. In 2007, the Port of Los Angeles opened the world's first alternative maritime power container terminal. The wharf at one berth was provided with the infrastructure to shore-power a compatible container ship. The electricity is converted to a voltage compatible with the ship's electrical requirements; the ship is connected to the shore power while hotelling at the dock. The ship shuts down its auxiliary generators.

There has also been some effort by cruise line companies to utilize this strategy while calling into ports. Princess Cruise Line, for example, utilizes cold ironing in Alaska and Seattle.

In general, the cost-effectiveness of cold ironing is dependent on several factors including the in-port electrical needs of the ship, how frequently a ship calls at a particular port, how long that ship is at port, and the energy demand required while at port.

Electrification of On-Road Vehicles at -

Many ports across the country have incorporated hybrid vehicles into their light-duty car and truck fleets. Hybrid vehicles utilize the dual technologies of gas and electricity. The wheels of the hybrid vehicle are driven by both a conventional engine that is fueled by gasoline or diesel fuel as well as an electric motor(s). The motors can be utilized in various ways, with a computer that is programmed to operate the vehicle on either or both the electric and ICE motors depending on the speed, the power required, and the amount of electricity left in the batteries.

There are few hybrid vehicle options for heavy trucks. Thus, for ports looking for electrification opportunities for heavy duty vehicles, hybrid technology is not currently a realistic strategy. There are, however, other electric strategies that involve heavy duty fleets at ports, including truck stop electrification and refrigeration units.

Electrification of the refrigerated truck, trailers and containers that are used by shippers to transport perishable items has significant electrification possibilities. There are many of these refrigerated units, generally called transport refrigeration units (TRUs), in the U.S. These TRUs may sit for hours, even days, at port terminals awaiting transfer to ship or truck. During this time, the refrigerated units are kept cold using an engine-powered refrigeration compressor typically run on diesel fuel. Grid-supplied electric power of these TRU engines is one strategy to reduce TRU emissions. This strategy involves the replacement of the diesel engine used to power the unit with an electric motor while the unit is idle at the port terminal.

Rail Electrification at Seaports

There are opportunities for electrification of the railway engines used by many ports to move cargo. In particular, switch engines—smaller locomotives used to move rail cars in a rail yard—may have potential for electric use due to their size and use specifications. There are two leading possibilities for electrification of switch engines:

- The all-battery engine, which is commonly used for smaller train sets or groups of rail cars of less than seven.

- The hybrid diesel-battery engine, which is capable of handling larger train sets of as many as 88-car mainline units.

Some smaller industrial facilities use rail movers, which are usually small rail-mounted vehicles powered by internal combustion engines to move single or small numbers of railcars. Larger train sets require larger traction power, and thus, hybrid technology is beneficial because idling is reduced and the diesel generation engines are much smaller than typical switch engines.

Both all-battery engines and hybrid diesel-battery engines result in substantial emission reductions. The all-battery unit achieves mobile source emissions reductions of 100% because it relies entirely on electricity, while the hybrid diesel battery technology achieves mobile source emissions reductions—NO_x and PM in particular—of 80 to 90% (EPRI 1002244). Total emissions must be compared to the conventional diesel engine(s) that is being replaced, the use of which may be highly variable. Some switch engines are heavily used while others see only periodical service.

RAILROAD ELECTRIFICATION

There are a number of existing options for moving people or goods via grid-connected electric rail. The principle technologies include (a) freight locomotives; (b) switcher locomotives; (c) regional passenger rail; (d) high-speed rail (rapid transit); and (e) light rail.

In the U.S., the majority of the fuel used in transportation is petroleum-based. Electricity comprises only a small fraction of total transportation energy use including rail transportation. There are currently no mainline freight locomotives powered by electricity operating in the U.S., and the only areas where significant portions of intercity railroads are electrified are rail corridors from Boston to Washington, DC; and Philadelphia to Harrisburg; as well as urban commuter rail lines. In contrast, many other countries (particularly in Europe and Asia) have railroad systems that are almost entirely electrified (EPRI 1020648).

Electric locomotives have a higher power-to-weight ratio than diesel counterparts, as on-board engines are not required for energy generation. Electric locomotives are faster and more powerful than diesels. Electric locomotives can accelerate and brake faster than diesels.

Electric locomotives also avoid many of the negative environmental effects associated with diesel locomotives. Powered by electricity generated miles from the train's operation, electric trains do not emit pollutants. This is particularly useful in operation in tunnels, indoor stations, rail yards, and underground locations. Emissions are also reduced overall, especially when electricity is produced from renewable or low-carbon energy sources. Electricity is also less expensive than petroleum when used as a transportation fuel, primarily due to the high efficiency of electric technologies.

According to a study done by Penn State University (Clemente, 2000), 1 Btu of electric rail is equivalent to 2.5 Btus of diesel rail. In other words, electric rail is 2.5 times as efficient as diesel rail. Likewise, 1 Btu of electric rail is equivalent to 17 Btus of freight hauled by diesel trucks. Penn State estimated that electrifying 20% of existing U.S. rail freight and transferring 30% of truck freight to rail would save over 1 million barrels of oil per day.

Electrifying railroads would consist of electric trains (non-Maglev) for coast-to-coast transport of people or goods. It would be a new market for electricity in the U.S. In this scheme, traction current converter plants would be used to convert grid electricity to an appropriate voltage, current type, and frequency to supply railways with traction current. Traction current converter plants are either decentralized (i.e., direct supply of the overhead line of the traction current converter plant with no feed into the traction current network) or centralized (i.e., for the supply of the traction power network and usually also for the supply of the overhead line). The availability of low-carbon electricity at the substations along the route could help enable this market. The environmental benefits of long-distance electric trains, particularly in the form of reduced CO$_2$ emissions, could be significant.

Transit systems such as metrolinks, light rail, subways, and monorails are used for public transportation. These systems typically use electricity from the grid. Often, electricity is supplied to the trains through overhead catenaries, with a return line in the tracks to complete the circuit. Electricity is a primary operational cost for electric transit systems. Electricity use in the entire transit sector was 5.649 million MWh in 2004. As carbon is "taxed"* access to low-carbon electricity could increase the market share and cost-effectiveness of electrically driven transit considerable. The most applicable forms of electric transit would be the larger subway and rail systems (e.g., the Bay Area Rapid Transit System around San Francisco, California called BART).

Locomotives for Goods Movement

Locomotives are usually used in specific applications (e.g., movement of freight or movement of passengers). Two types of locomotives used in freight transport are (1) freight locomotives and (2) switch locomotives. Electrification of these types of locomotives is possible through powering from an external source of electricity (via overhead lines or third rail) or through use of batteries. Electric locomotives are distin-

---

*The author suggests that some form of carbon tax is highly likely in the decades ahead.

guished from diesel-electric locomotives, which are powered by diesel engines and use electricity in only the traction. system.

- Freight locomotives are used to haul thousands of tons of freight including raw materials and finished goods. Freight locomotives weigh hundreds of tons and have high horsepower (HP) ratings (~4,000 HP and up). In the U.S., freight locomotives are almost exclusively powered by diesel engines. Their fuel consumption can be significant: a 4,000-HP freight locomotive can consume up to 500,000 gallons of diesel fuel each year (CARB,2009). There are currently no electric locomotives used in mainline freight services in the U.S.

- Switcher locomotives ("switchers") assemble and disassemble trains and move railcars within a rail yard. Switchers have a power range lower than freight locomotives, typically 1,000 to 2,300-HP rating. In contrast to freight locomotives, switchers spend a large portion of time in lower-power or idling modes; an average switch locomotive may consume about 140 gallons of fuel per day, or 50,000 gallons each year (CARB, 2009). Switch locomotives in the U.S. are almost exclusively diesel-powered.

Locomotives for Passenger Transport

Regional passenger rail, or commuter trains, are powered by locomotives or can be powered by individual cars known as electric multiple units (EMUs) which include propulsion in the passenger car. Commuter trains transport passengers between city centers and suburbs with speeds varying between 30 to 125 miles per hour. There are a number of rail systems operating in large cities in the U.S. including New York City, Chicago, San Francisco, and Washington, DC. Electrification of passenger rail is possible and common compared to freight locomotives. For example, the majority of the Northeast Corridor (a 600-mile system between Boston and Washington) includes overhead electrified wire. However, electrification of regional rail is largely lacking in the remainder of the U.S.

High-Speed Rail

High-speed rail involves transportation by rail sustaining speeds over 125 miles per hour, usually over longer distances. Given safety considerations from shared right of way with freight trains and slower passenger trains, U.S. Federal Railway Administration regulations limit train speeds; high-speed rail is, therefore, rare in the U.S. In Europe and Asia, high-speed trains are more common.

Light Rail Vehicles

Light rail vehicles transport smaller volumes of passengers over shorter distances at low speeds. Almost exclusively powered by electricity, light rail vehicles are usually made up of EMUs. Light rail systems are used for travel within urban areas. There are a number of systems in place in cities like San Francisco, San Jose, Seattle, Washington, DC, and many others.

Maglev (Magnetic Levitation)

Maglev is an emerging application in which trains are levitated above guide rails by electromagnetic forces. The trains are equipped with lightweight, powerful magnets, and the guide rails contain electrified coils. The application is very electricity-intensive. Magnetic fields are created in cables or on the vehicle itself, and the magnetic fields propel the vehicle (CARB, 2009). The very few existing Maglev systems are used for passenger transport, although it is possible to produce freight vehicles that carry freight. Maglev trains are currently in the early commercialization stage in Japan and China. Low-carbon electricity could help Maglev trains achieve greater market penetration. Implementation of Maglev technology has been extremely limited due to very high infrastructure costs, small marginal speed benefits in short hauls compared to alternative options, and public concern about exposure to electromagnetic fields.

Linear Induction Motors

Linear induction motors (LIMs) are an emerging electrification technology. LIM technology is used to push the train along the track with a magnetic field. LIMs use electrical current to generate the magnetic field which propels the train. LIMs can be used with steel wheel on steel rail. Several passenger transport systems of a few miles in length exist (CARB, 2009). LIM infrastructure is expensive compared to conventional rail, and the feasibility for longer rail lines or freight transport is uncertain.

MINING ELECTRIFICATION

Electricity is used to power much of the machinery operating in mines. Longwall and continuous mining operations in underground mines rely on electricity supplied from above ground or from underground electrical substations. In surface mines, the largest draglines and drills are also electrically powered. Many mining operations, however, still rely on diesel-powered heavy equipment and could benefit from electrification. Electric drives are more cost-effective, reliable, operationally efficient, and environmentally superior. They enhance worker health and safety as well (EPRI 1020278).

Electric mining equipment options include conveyance systems, shuttle cars, ram cars, haulage systems, draglines, electric shovels, rail locomotives, personal vehicles, forklifts and other equipment.

Conveyance Systems

In many mining applications, electrically powered conveyors can replace internal combustion vehicles that transport materials and perform a series of loading and unloading handoffs. Factors that determine feasibility, cost, and payback of such systems include distance or horizontal length of the conveyance system, vertical life, maximum size, and composition of the material.

Shuttle Cars, Ram Cars, Haulage Systems

Whether the application is surface or underground mining, haulage vehicles transport the mined material from its origin to a handoff point. These vehicles are made with both electric and diesel drive trains by a number of manufacturers and can be operated by a driver or by remote control. In many cases, heavy-duty diesel trucks could be replaced with electric shuttle or haul cars that can do the same job more efficiently, less expensively, and in a more environmentally beneficial manner.

Some manufacturers offer AC-powered electric drive systems, which significantly increase a vehicle's traction. AC drive motors require less maintenance than DC-powered systems because they have no brushes to inspect and replace.

Draglines and Electric Shovels

Used in surface mining operations, draglines are machines with a scoop suspended from a crane and maneuvered by a series of ropes and chains. Among the largest mobile machines on earth, draglines remove dirt to expose coal or minerals. Most mining draglines are all-electric, powered by a trailing cable connected to a medium-voltage DC source or land barge that serves as a step-down power source.

Rail Locomotives

As discussed previously, battery-powered or trolley-style electric rail locomotives offer another electric alternative to diesel locomotives.

Personnel Vehicles

Personnel vehicles, which transport miners and equipment from the surface of the mine to the underground working locations, come in battery-powered electric and internal combustion options.

Forklifts and Custom Material Handling Equipment

Battery-powered forklifts are commonly chosen over internal combustion forklifts in interior warehouse and industrial settings.

NEW OPPORTUNITIES FOR ELECTRIFICATION

There are a few opportunities for electrification which have evolved, in part, as a result of the shift from fossil fuels to electric energy in a number of applications.

Hydrogen Production Plants for Fueling Vehicles

This application involves the production of hydrogen for use as a fuel in vehicles (including cars, trucks, forklifts, etc.). The hydrogen could be applied either in fuel cells or in modified internal combustion engines. Since use of vehicles is generally dispersed, this application would require hydrogen generation at a central location and a distribution infrastructure. More than 80% of all hydrogen produced worldwide is generated by steam methane reforming of natural gas, since this is presently the lowest cost process for large-scale production (~$1.5/kg H2 or ~$0.7/lb H2). Currently, most hydrogen is used by the chemical industry. However, application of hydrogen to the transportation sector is under active investigation, in large part because of its environmental benefits. The availability of a low-cost hydrogen generation process and distribution infrastructure could help propel hydrogen's use as a fossil-fuel alternative.

Military Bases

Military bases use energy for launching aircraft, rockets, missiles, and other projectiles, as well as for general base operations. They also use fuel in airborne and land-based vehicles. Several of these applications now use inefficient fossil-fueled generators. Electricity can be used to produce hydrogen for fueling aircraft, spacecraft or other vehicles. The U.S. Air Force is actively implementing energy-efficient measures, investigating alternative jet fuels, and pursuing green energy technologies. It is the largest purchaser of renewable energy in the U.S. (buying more that a million MWs in 2005); in addition, 25% of its vehicle fleet consists of flexible fuel vehicles. Because of their interest in demonstrating leadership in the implementation of environmentally friendly energy solutions, they would be a prime candidate for electrification.

References

"Airport Emissions Quantification—Impacts of Electrification," L. Sandell, EPRI, Palo Alto, CA: 1998. 111000.

"Non-Road Electric Vehicle Emissions—Analysis and Recommendations," R. Graham, EPRI, Palo Alto, CA: 2003. 1002244.

"Seaport Land-Side Equipment Electrification Opportunities—At the Port of Houston, Texas, the Port of Long Beach, California, and the Port of Toledo, Ohio," A. Rogers, EPRI, Palo Alto, CA: 2006. 1010577.

Memorandum: "Off-Road Electrification Technical Analysis: Screening Assessment Results," Eladio Knipping, EPRI, August 9, 2010.

"Electric Ship to Shore Cranes: Costs and Benefits," EPRI, Palo Alto, CA: 2009. 1020510.

"Electric Retrofit for Airport Container Pallet Loader," EPRI, Palo Alto, CA: 2009. 1020482.

"Electric Outdoor Lift Trucks: New Technologies Meet Customer Needs," EPRI, Palo Alto, CA: 2009. 1020515.

"Mining Electrification: Potential and Benefits," EPRI, Palo Alto, CA: 2009. 1020278.

"Plug-In Hybrid Yard Tractor: Demonstration Plan," EPRI, Palo Alto; CNP, Houston, TX; NYPA, New York, NY; SCE, Rosemead, CA; Southern Company, Birmingham, AL: 2009. 1017678.

"Options and Opportunities for Rail Electrification in the United States," EPRI, Palo Alto, CA: 2010. 1020648.

"Technical Options to Achieve Additional Emissions and Risk Reductions from California Locomotives and Railyards," California Air Resources Board, CARB, August 2009.

"Electricity is the means to a better life across the world," F. Clemente, Penn State University, 2000, http://www.coalcandothat.com/assets/resources/Monday%20Beneficial%20Electrification_v3.pdf

Chapter 6

Beneficial Industrial Uses of Electricity: Industrial Introduction and Process Industries

There are many opportunities for beneficial electricity in the industrial sector. Electric technologies in industrial processes use electricity to make or transform a product. Many electrotechnologies are used for heating applications including heat treating, drying, curing, melting, and forming. Others are used for applications like motive power, separation, machining, and welding.

Electric technologies have good controllability, superior product quality, cleanliness, and efficiency. In many cases, electrotechnologies are chosen for technical reasons, while in other cases, the relative price of natural gas (or other fuels) as compared to electricity is the deciding factor. In a number of cases, the application cannot be done effectively without an electrotechnology.

Electrotechnologies are adopted within industries because of their efficiencies, energy control capabilities, low material waste, and high processing and production rates. The fact that they reduce CO_2 emissions as well will increasingly provide an added incentive for their development and use. This will be particularly true if and when CO_2 "taxes" are in place.

There are a number of electrotechnologies which are not fully deployed which reduce or eliminate other adverse environmental impacts often associated with gas-fired processes. For example, drying automobile paints with gas ovens releases volatile organic compounds (VOCs) into the atmosphere. New powdered paints, dried using electrotechnology, eliminate these hazards and produce a superior finish.

In the industrial (and municipal) markets, the practical technical potential of beneficial electrification consist of a variety of electrotechnologies. In the industrial market, the technologies with the greatest potential include induction heating, direct arc melting, freeze concentration, and electrolytic separation. In municipal water, waste water and waste applications, electrotechnologies with the greatest practical technical potential include membranes for desalination, ozone or ultraviolet disinfection and purification of water, and vitrification of industry waste. Table 6-1 lists electrotechnologies by industry category.

Figure 6-1 pictures examples of technology applications which reduce CO_2 by using electricity to displace fossil fuels.

Table 6-1. Electrotechnologies by Industry Category (Sources: EPRI CU-7441 and 1013998)

Process Industries

    Electrochemical Synthesis
    Electrolytic Separation
    Freeze Concentration
    Industrial Process Heat Pumps
    Membrane Process

Metals Production

    Direct Arc Melting (Arc Furnaces)
    Electrogalvanization
    Electrolytic Reduction
    Electro Slag Processing/Remelting
    Induction Heating and Melting
    Ladle Refining
    Plasma Processing
    Vacuum Melting Automation
    Induction Heating
    Infrared Processing (Electric)
    Laser Processing
    Microwave Heating, Drying, and Processing
    Radio-Frequency Heating and Drying
    Resistance Heating and Melting (Direct and Indirect)
    Ultraviolet Curing

Materials Fabrication

    Electric Discharge Machining
    Electrochemical Machining
    Electrofinishing
    Electroforming
    Electron Beam Processing
    Flexible Manufacturing Systems

Industrial Heat Pumps

Process Cooling & Refrigeration

Electric Process Heating

Heat Recovery

Electrolytic Reduction

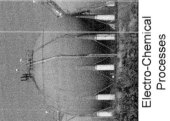

Electro-Chemical Processes

Figure 6-1. Examples of Technologies Applications Which Reduce CO_2

PROCESS INDUSTRIES

Process industries comprise a mixture of widely disparate manufacturing markets including the chemicals, paper, food, petroleum, textile, and tobacco industries. These manufacturing markets range from the highly dynamic organic chemicals and biotechnologies industries to the mature tobacco industry. The unifying theme is the manufacturing process involved, namely the almost continuous and prolonged processes and their reliance on the following five electrotechnologies:

- Electrolytic separation
- Electrochemical synthesis
- Freeze concentration
- Industrial process heat pumps
- Membrane processes

Electric-based process heating systems use electric energy or electromagnetic fields created by electricity to heat materials. Direct heating methods generate heat within the work piece by either:

1. Passing an electrical current through the material,
2. Inducing an electrical current into the material, or
3. By exciting atoms and molecules within the material with electromagnetic radiation.

Indirect heating methods use one of these three methods to heat an element which transfers the heat either by conduction, convection, radiation or a combination of these to the work piece.

The following are examples of process industries.

Papermaking
The papermaking process can be divided into three basic steps: Stock preparation, papermaking itself, and finishing. Stock preparation involves the chemical and mechanical treatment of a fiber/water slurry originating from a mill's own pulping operations or in dry 500- to 600-pound bales when purchased from outside. All pulps are diluted to about the same consistency before the slurry is mechanically refined to prepare the fiber for the paper machine. Stock refining is usually done

in large-horsepower equipment that, through shear stresses between rotating plates, alters the individual fiber's structural characteristics. Depending on the paper or board grade to be manufactured, the refining energy requirements can vary from almost nothing to nearly 250 kWh per ton of paper (only small quantities of specialty papers require considerable higher refining energy inputs).

Before entering the paper machine, fibers are also typically cleaned with centrifugal cleaners and mechanical screens. Entrained air in the stock is removed from the fiber water slurry just before it enters the paper machine. The temperatures of the fiber/water slurry in stock preparation are elevated and kept constant with steam to improve water removal on the paper machine. Paper machine configurations begin with a headbox(es). The headbox is used to form a thin, even layer of the low-consistency stock that is spread out on or between rotating wires.

All other paper machine components are designed to remove water or surface treat the paper or paper-board after the stock layer is formed. Although there are many variations, the following three stages of water removal are typically used. In the first stage after the headbox, the dilute suspension of mainly fibers and sometimes mineral fillers is subjected to water drainage through one or more porous continuously rotating belts called wires. In the second phase of the water removal, pressing, the fibrous web is pressed in up to four nips to remove water mechanically. Sometimes the web is heated with direct steam to aid water removal during this stage. The third stage of water removal is drying in which the remaining water is removed through an evaporative drying process. The amount of water removed decreases in each successive stage, and the costs of removing the water increase considerably with each stage. At the dry end of the paper machine, the moisture content is typically around 4 to 8%—close to the natural moisture content of paper or board stock in ambient air. The dry paper is wound continuously onto a reel spool the same width as the paper machine.

Food Processing

Food processing is one of the most ubiquitous industries in the world. Food is processed in every state and in every country. Processing encompasses the conversion of raw ingredients, primarily from agricultural sources to shelf-stable, convenient-to-use forms in food service or retail-sized packaging with consumer-friendly attributes. This includes:

- Meat products
- Dairy products
- Preserved fruits and vegetables
- Grain mill products
- Bakery products
- Sugar and confectionery products
- Fats and oils
- Beverages

Methods vary from simpler ones such as the washing, sorting, packaging and refrigeration of fresh produce, to the more complex ones of formulating, heat processing, multi-component assembling, and freezing of products such as ready-to-eat dinners, entrees or desserts. Processes used by the food industry to impart maximum practical shelf life to food products include:

- Aseptic processing
- Canning
- Dehydration or partial dehydration
- Fermentation
- Freezing and refrigeration
- Irradiation
- Non-thermal processes
  — Pulsed-electric pasteurization
  — High-pressure sterilization
- Pasteurization

Food Processing Innovations
An EPRI study (EPRI CU-6755) identified these electrotechnologies as the leading food processing innovations in the 20th century:

- Aseptic process
- Microwave oven
- Frozen juice concentration (freeze concentration)
- Ultra-high-temperature short-time sterilization

Electricity plays a major role in the successful application of each of these innovations by supplying energy, facilitating control, and providing accurate monitoring. Opportunities for the application of electrotechnology exist in many of the food manufacturing processes. For example, microwave heating has been shown to be a rapid, economical method of defrosting ingredients with resulting higher yields while providing better microbiological control. Other examples include:

- **Irradiation** promises to be a major factor in improving food safety, albeit it, some issues of consumer concern remain. The rapid and effective reaction of ozone with microorganisms, coupled with its excellent odor and flavor characteristics, has multiple applications as a sanitizer and disinfectant for the food industry.

- **Ohmic heating** offers many advantages for aseptic processing and other process heating applications because of its ability to heat pumpable materials rapidly without the use of hot metal surfaces such as those found in conventional heat exchangers.

- **Freeze concentration** is an excellent technique for removing water from products such as fruit juices and dairy products where the application of heat or vacuum can be very detrimental to delicate flavor notes or to the product color.

- The multiple applications of **membrane technology** for process separation and/or concentration, the recycling of process streams, and the clean up of effluent streams have only begun to be realized.

Electrotechnologies in food processing can improve productivity and reduce processing costs. Examples include energy integration and efficiency with the use of heat pumps, direct heating with microwaves or ohmic heating, and concentration without the application of heat such as the above-mentioned freeze concentration or membrane separation.

An Example of Food Processing—Breakfast Cereals

Processing breakfast cereals is a capital-intensive, multi-stepped, time-consuming operation. Traditional processes require 30 to 72 hours from the time grains are received in the plant to the time the finished product is ready for packaging.

Breakfast cereals production commences with grain preparation, a process in which the grain (oats, corn, wheat, or other grains) is processed to enable additives to be easily absorbed and to render the grain suitable for forming (Figure 6-2). Large horsepower motors are used to grind, chop, and clean huge quantities of grain. The prepared grain is then fortified with nutritional supplements, such as vitamins, minerals, or desired flavors. Once the proper mix is achieved, it must be cooked at low temperatures (approximately 120 to 200°F) for one to two hours. The heated grain is then flash-dried, slightly cooled, and hammer-milled to separate lumpy material. These steps take anywhere from six to twelve hours. The cooked and separated grain mix is now ready to be formed into final products. The mix is forced through sized die casts or slicing mechanisms to slice, shape, or shred the cereal. Depending on the end product desired, the shaped grain is covered with sugar, syrup, spices, or other coatings. The cereal product is then toasted, baked, or dried to ensure that it has the proper texture, flavor, and moisture content. The ready-to-eat cereal is finally packaged for distribution.

Irradiation of Food Products

Major improvements in food safety are possible with the use of ozone and the application of irradiation. Food irradiation is the controlled exposure of food, either before or after packaging, to a source of high-energy invisible ionizing radiation from Cobalt 60 or from electrically generated electron beams. Irradiation has a lethal effect on food pathogens, those microorganisms that cause food-borne illness, thus it can be used very effectively to improve food safety and reduce the possibility of food poisoning.

Radiation is not a new man-made innovation. Natural radiation reaches us daily from the sun and from other sources in our environment. Even the conventional baking of food in an oven exposes the food to infrared radiation. Both conventional baking and controlled irradiation cause similar harmless chemical reactions in food.

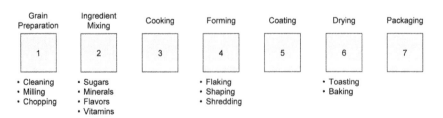

Figure 6-2. Basic Steps in Breakfast Cereals Production

Radiation has been studied in the U.S. and elsewhere ever since radioactive materials and X rays were discovered in the 1890s. Research has addressed the biological effects of radiation, the safeguards against potential hazards, and the multiple beneficial uses. The first patent application for preserving food by irradiation was filed in 1905.

Irradiation of food products has been accepted as a safe and effective method by the World Health Organization, the Food and Agriculture Organization, American Medical Association, American Public Health Association, Institute of Food Technologists, American Society for Microbiology, and the Food and Drug Administration (FDA) through specific product approvals.

Irradiation can be used to:

• Improve food safety by killing any illness-causing microorganisms present in a food product.

• Destroy insects and insect eggs thus negating the need for chemical fumigants.

• Destroy spoilage organisms resulting in extended shelf life for fresh produce and fresh fish, meat or poultry.

• Inhibit the sprouting of root crops such as potatoes and onions thus eliminating the need to apply chemicals.

• Slow the ripening of fruit.

Irradiation damages or destroys the structure of the membranes around the cells of all types of bacteria including the microorganisms (food pathogens) such as Salmonella, Campylobacter. E. coli 0157:H7, and Listeria monocytogenes that cause food poisoning.

Irradiation also causes irreversible changes in the DNA of the living bacterial cells. These changes are caused by the reaction of DNA with the short-lived freed radicals and peroxides formed when food products are exposed to irradiation. Since the food pathogens are unable to reproduce after exposure to irradiation, they cannot multiply and the food products are, in effect, free of potential illness-causing microorganisms.

After exposure to approved levels of irradiation, any toxin-producing microorganisms are no longer capable of producing more toxins. However, the irradiation does not significantly decrease the levels of any toxins that had been formed prior to irradiation. A major advantage of the use of irradiation is the ability to destroy pathogens in foods without affecting the original flavor, odor, color and texture of most food products.

Irradiation has been successfully used for microbial control on a wide variety of food products including:

- Grains and cereal products
- Fruits and vegetables
- Prepared foods, e.g., salads, cole slaw, and entrees
- Dairy products including milk and cheese
- Fresh and frozen meat, especially hamburger, plus poultry and seafood
- Processed meats, e.g., frankfurters, sausage and bacon
- Beverages such as wine and fruit juices
- Seasonings, herbs and spices

Irradiation Sources

There are many types of radiation and radiant energy. They differ primarily by their wavelength and frequency. The shorter the wavelength, the greater the amount of energy per unit. Starting with the short wavelength sources in this electromagnetic spectrum, there are:

- Cosmic rays which come from sources in interstellar space
- Gamma rays which are emitted from radioactive atoms
- X-rays which are generated by X-ray tubes
- Ultra-violet rays from the sun and from sun lamps
- Visible light from the sun, light bulbs, fire, etc.
- Electron beams
- Infrared rays from the sun, oven heat elements, and sun lamps
- Radar and microwaves which are generated by their respective sources
- TV, both UHF and VHF sources
- Standard radio broadcast signals
- Electric waves
- Standard 60-cycle household electricity

The type of irradiation used to treat food products is just one seg-ment or slice of this continuum of electromagnetic radiation. There are three energy sources that can be used for the irradiation of foods, X-rays, electron beams and gamma rays. In practice, two sources are used, electron beams and gamma rays from Cobalt 60.

Electron Beams

Electrically powered electron beam (E-Beam) accelerators capable of continuous operation at 5 MeV or 10 MeV produce electron beams. Electrons generated by a cathode are accelerated down a tube and fo-cused on a metal target such as tungsten foil. When the electrons strike the tungsten, they suddenly decelerate and some of the electrons are converted to electromagnetic energy (photons).

In order to improve the uniformity of penetration and dosage, the lower-energy photons in the electron beam are filtered out using suit-able shielding materials. The residual beam is then directed at the food product or other material to be irradiated.

The advantages of the electrically generated radiation include:

- No radioactive materials are stored on site (eases concerns of workers and local residents).

- Eliminating security issues and security costs associated with the storage and transportation of radioactive materials.

- Radiation emissions stop immediately when power is turned off, an obvious safety advantage.

- Amount of irradiation can be adjusted more easily than with a Co-balt source.

- Fewer government permits are required to build and operate a fa-cility.

The primary disadvantage of the E-Beam is the fact that E-Beam-generated radiation does not penetrate a food product as deeply as does radiation from a Cobalt 60 source. Whole pallets of product can be pro-cessed in a Cobalt 60 facility while irradiation treatment using an E-Beam generator is typically limited to single layer packages.

Gamma Rays

Gamma rays for use in food irradiation are produced by the decay of Cobalt 60. While Cesium 137 can also be used, the long-term uncertainties concerning a continual supply of Cesium 137 has caused industry to phase it out as a radiation source.

The production of gamma rays by Cobalt 60 is a natural decay process. It is not a nuclear reaction, and there is no danger of a "meltdown," an explosion or a chain reaction occurring. Cobalt 60 cannot make materials radioactive since it does not produce neutrons. There have been over one million shipments of Cobalt 60 in the U.S. since 1955 without a single radiation accident.

Radioactive waste does not accumulate at irradiation facilities since no radioactivity is produced. The Cobalt 60 source has to be replaced after its ability to produce irradiation has decayed significantly below its original level—usually in 16 to 21 years. The decayed Cobalt 60 source is then returned to the supplier who can reactivate the decayed materials or store it in the prescribed manner.

In a gamma ray irradiation facility, the Cobalt 60 is in slender pencil-like stainless steel casings that are contained in a lead-lined chamber. The packaged food travels in pallets on a conveyor between 6.5-foot thick concrete walls into and through the irradiation chamber where it is exposed to the gamma rays. The amount of irradiation treatment is controlled by the length of time the pallet is in the irradiation chamber.

Irradiators are designed with redundant safety systems to ensure no incidence of over-treatment or exposure of workers to irradiation sources. Potentially hazardous areas are monitored and alarmed, and systems of interlocks are in place to prevent unauthorized entry to an irradiation area. Strict operating procedures must be enforced, and worker training must be thorough with periodic refresher sessions.

Aseptic Processing

A shelf-stable product must be safe and appetizing to eat after extended storage at ambient or above-ambient temperatures. To meet these minimum criteria, the food product must be commercially sterile; i.e., must be free of the common food pathogens, those microorganisms that cause illness when consumed. A shelf-stable product must also be free of spoilage organisms, such as yeast and molds, and must have minimal retained enzymatic activity.

Common food pathogens can be destroyed by heating the food

product to a given temperature and by holding the product at that temperature for a specified period of time. The same time/temperature conditions are usually sufficient to destroy most spoilage organisms and to inactivate most enzymes.

In aseptic processing, the product is sterilized before packaging by rapid heating, using one of several approved methods including a scraped surface heat exchanger, plate or tube heat exchanger, or by ohmic heating. The sterilized product is rapidly cooled, then held in a sterile environment until deposited in a pre-sterilized container. While still in a sterile environment, the package is sealed, and only then is it discharged to ambient conditions.

Heating and cooling the product prior to packaging allows the product to be rapidly heated and cooled, thus minimizing heat degradation of the texture, flavor or color. Comparative data from the Aseptic Packaging Council (www.preparedfoods.com) shows typical processing times (heating, holding, cooling) for an acidic low-viscosity product:

- Aseptic Process: 3 to 15 seconds
- Hot Fill Canned: 8 to 15 minutes
- Retort Canned: 20 to 50 minutes

Obviously the reduced heat exposure time of the aseptically processed product makes a marked difference on the relative quality of the aseptic versus the canned product. Since aseptically processed products are cooled prior to packaging, the possibility of extended heat retention of cartooned and palletized products with resultant product degradation, often referred to as "stack burn," is eliminated.

Sterilization prior to packaging means that the degree of heat treatment used for small retail packages is the same as that used for larger food service-sizes containers or bulk containers. Thus, the product quality in a bulk container or a food service-sized container can be identical to that in a small retail-sized packages. This is definitely not the case for canned product where food service-sized cans must be heat processed significantly longer than retail-sized cans with the resultant increased degradation of the product quality in the larger containers.

In most aseptic packages, the product does not touch a metal surface so there is no transfer of metallic flavor as is often the case with juices or fruits packed in tin-coated steel cans.

Other Food Processing Electrification Opportunities
Wet Corn Milling Facilities: Ethanol Production
through Fermentation of Grains

Wet corn milling is the most energy-intensive industry within the U.S. food and kindred products group, using approximately 15% of the energy in the entire food industry. Electricity is used primarily for machine drive (pumping, grinding, separating, and drying). Steam is used for evaporation, drying, fermentation, extraction, and ethanol recovery. Drying accounts for the majority of energy consumption. A 100,000 bushel per day wet mill uses ~320 MWh of electricity and ~760 MWh of steam. A typical wet corn milling plant spends approximately $20 to $30 million per year on energy. The energy consumption per dollar of value added is relatively high (0.019 MWh).

Ethanol is produced from fermentation of grains. Corn accounts for approximately 95% of U.S. ethanol production. The other 5% is produced from feed stocks such as sugarcane; barley, wheat and sorghum grains; various types of cellulosic biomass; and certain types of industrial waste. In 2004, ~75% of U.S. corn-based ethanol was produced in dry corn mills and ~25% was produced in wet corn mills. The wet mills are larger and produce more valuable byproducts than dry mills. A typical 24.5 million gallon (93 million liter) wet corn mill ethanol plant spends $0.15 on energy (power, heat, and cooling) for each gallon ($0.04 per liter) of ethanol produced.

Ethanol production has skyrocketed in recent years, from ~175 million gallons (~600 million liters) in 1960 to ~3.4 billion gallons (~13 billion liters) in 2004 due to government mandates. (Note that if hydrogen can be produced and distributed in a cost-effective manner in the future, the market share of ethanol as a fuel will likely decline.)

The availability of low-carbon electricity could considerably lower operating costs at wet corn milling facilities. One significant new use of electricity could be to replace fossil-fuel-fired boilers with electric boilers (EPRI 1014570).

Rendering Facilities: Biodiesel Production
through Transesterification of Waste Oils and Fats

Rendering facilities, which produce potential waste feed stock (yellow grease, lard, and tallow) for the biodiesel industry, employ very energy-intensive processes. Fuel to operate rendering facilities (e.g., heat and steam to cook the animal wastes) accounts for ~30 to 35% of operating costs. Electricity (e.g., for grinding) accounts for another 10 to 15%. Thus, energy represents nearly 50% of operating costs. Access to low-carbon electricity could also be instrumental in the realization of biodiesel production using waste grease and oil. Rendering facilities could accept recycled grease and trap grease from food manufacturers and restaurants. Currently, there are documented tipping fees of up to $0.11 per gallon ($0.03 per liter) for urban waste grease such as recycled cooking oil and trap grease (EPRI 1014570).

MEMBRANE PROCESSING

Membrane processing depends on an electric motor-driven pump to drive a product stream through a semi-permeable barrier for purification and to fractionate or concentrate liquids and gases. Membrane separation technologies include reverse osmosis, ultra-filtration, and microfiltration. This technology has advantages over technologies such as vaporization which include lower energy use, better production reliability, reduced footprint, lower capital costs, lower emissions, and better product quality. In addition, salable byproducts such as corn syrup and edible oils can often be reclaimed from waste streams. Membrane processing is well-suited for processing businesses that handle fluids— from wine to cheese, juice to gelatin, for fluid clarification, concentration, water filtration, waste treatment, and more.

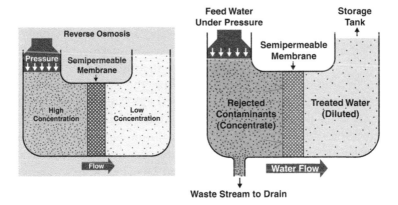

Figure 6-3. Membrane Application Example: Reverse Osmosis

In 2001, the food and beverage processing industry accounted for 21% of the membrane market (following water and wastewater treatment, which accounted for 57%) (Fredonia, 2002). In many industrial sectors, membrane technology has the potential to handle the majority of water treatment needs. Though reverse osmosis (RO) is currently the best available membrane technology for water treatment because of its ability to remove most impurities from water, microfiltration still accounts for more than half of the share of membrane processes. However, the demand for finer membranes such as nanofiltration and RO is projected to increase sharply in the near future. Additionally, specialized membranes are being developed to serve as stand-alone treatment technologies or to work in concert with other treatment technologies.

Other Industrial Applications for Membranes

Membrane separation provides industrial operators with the ability to separate extremely small diameter solid particles from liquids, to "filter" out specific solutes from a solution, and to separate one liquid from another by the use of sophisticated semi-porous ceramic, sintered metal, or polymetric film membranes. Membrane separation can be considered a modern, highly efficient, versatile form of the traditional filtering process. The objective is still the same: removing one component, such as a solid, from a second component, such as a liquid, or separating a given molecular component from a solution.

What makes membrane separation so noteworthy is the efficiency with which separations can be made and the wide range of components that can be processed. Separations can be done based on particulate size differences of less than a micron (0.00004-inch). Separation depends on the passage of specific molecules through a semi-permeable membrane while other molecules are retained by the membrane.

In traditional simple filtration setups, the osmotic pressure due to differences in concentration on the two sides of a membrane and/or gravity were the only driving forces. Modern membrane separation systems normally apply pressure to the solution being filtered to speed up the separation process. In some instances, when highly ionic systems are being filtered, an electrical potential will be applied to accelerate the process.

Membrane separation is typically broken down into four sub-technologies based on pore size of the filter component, namely:

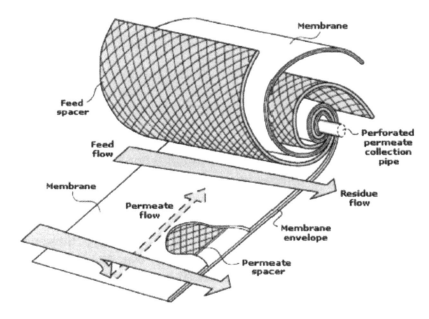

Figure 6-4. Schematic of a Spiral-Wound Membrane (Global Energy Partners, 2009)

- **Microfiltration**: Microfiltration is the coarsest of the four membrane technologies. It is applicable for separating suspended solids with particulate diameters greater than 0.1 microns. The pressure drop across the membrane ranges from 1 to 25 psig.

- **Ultrafiltration**: Ultrafiltration is used to remove suspended particles with diameters of less than 0.1 microns. The pressure drop across the membrane is about four times greater than with microfiltration and, as expected, throughputs are lower. Ultrafiltration retains substances with molecular weights about 1000 to 500,00 daltons.*

- **Nanofiltration**: Nanofiltration, with its smaller pore size, can be used to retain molecules with molecular weights above 100 to 1000 daltons. The pressure drop across the membrane is often more than double that of ultrafiltration. One of the main advantages of

---

*A dalton is a unit of atomic mass roughly equal to the mass of a hydrogen atom (1.67 x 10^{-24} g).

nanofiltration is that it can hold back molecules like sugars while letting salts pass through.

•   **Reverse Osmosis**: Reverse osmosis (RO) uses the smallest pore sizes of the four membrane separation processes and can be used to concentrate most water-based solutions. RO retains substances with molecular weights above 50 to 100 daltons, while smaller molecules (such as water with a molecular weight of 18 daltons) pass through the filter. The pressure drop across the filter is 20 to 100 times greater than with microfiltration.

Typically, the suitable choice of the membrane and the type of filtration to use is made on a case-by-case basis. To be effective, the selected membrane must have pore diameters that will allow one component to flow through the pores in the membrane, while preventing the other component from passing through. It is very important to choose the correct type of membrane in order to optimize and maintain the flow rate. Therefore, a membrane with the maximum pore size consistent with providing the desired separation is usually selected.

Membrane technology can be considered mature. It has been used commercially in the U.S. for more than 70 years, with its first applications in the water, wastewater, and beverage industries. However, the penetration of industrial membranes is expanding rapidly as a result of technological advances, and membrane technologies are used today in the water, wastewater, beverage, and food industries, as well as the drug and medical markets. The development of more advanced membrane materials (particularly polymers, ceramics, and sintered metals) has led to lower costs and greater versatility. Newer membranes have increases durability, improved efficiency, and higher capacity and are less prone to clogging than early membranes.

As a result of its maturity, there are numerous other industrial applications of membrane separation technology, as illustrated in Table 6-2. Common industrial processes benefiting from membranes include recovery of waste products, concentration, purification, and filtration. A variety of industries currently employ membrane systems. In particular, they are used extensively in the food, dairy, beverage, chemicals, and petroleum industries.

Table 6-2. Examples of Typical Industrial Membrane Applications (Source: EPRI 1020680)

Industry	Recovery	Concentration	Purification/Filtration
Food Processing	Recovery of lactose and whey protein from salty wastewater Recovery of caustic cleaning solution Recovery of acid cleaning solution Recovery of sugar from rinse water	Preconcentration of dilute sugar Concentration of fruit juices, juice flavor, syrup concentration, starch water, steep water, etc. Concentration of starch wash water Fractionation of steep water Enrichment of dextrose	Purification of process water Removal of pulp
Dairy	Recovery of Clean-in-Place solutions	Concentration of cheese whey, milk, lactose, egg, etc. Casein fractionation	Wine and beer filtration Fat removal Desalting of sat whey Bacteria removal Removal of color from wine Removal of alcohol from wine and beer
Beverage	Recovery of caustic solutions used for cleaning PET		Purification of process water
Pulp & Paper	Recovery of paper coating	Concentrating solid waste products	
Textile	Recovery of textile size	Desalting of dyes	
Chemicals & Petroleum		Gaseous separation Preconcentration of evaporator feedstock	
Pharmaceutical	Recovery of Clean-in-Place solutions	Production of desalted, concentrated antibiotics	
Automotive	Recovery of electrocoat paint		
Electronics			Purification of process water
Most Industries	Recovery of waste products from wastewater		Treatment of captured process water for reuse (e.g., reuse as boiler makeup feed water) Treatment of wastewater

FREEZE CONCENTRATION

Freeze concentration is also a relatively low-cost method of concentrating water-based solutions without the application of heat. The technology offers the food processor a mechanism to produce high-quality, heat-sensitive products, such as fruit juice and dairy concentrates, without loss of their original flavor, color and nutritional properties.

Freeze concentration has been used commercially since the 1950s with one of the first applications being in the polyester industry. Starting in the early 1980s, freeze concentration was being used for the production of fruit juice concentrates and for the concentration of coffee. By the late 1990s, there were reportedly over 70 commercial freeze concentration systems in operation worldwide.

Theoretically, freeze concentration is applicable for the removal of water from any fluid product and is especially advantageous to those that are heat sensitive. Food industry applications include:

- Juice concentrate production
- Desalination of water
- Beer and wine production
- Concentration of coffee
- Milk and whey concentration
- Concentration of vinegar
- Removal of impurities from wastewater streams

Advantages over conventional concentration methods include:

- No heat degradation of flavor and color
- No development of burn-on nor black specks
- Reduced chemical and cleaning costs
- Extended run times with resultant efficiency improvements
- Potential energy savings
- Higher yields

Freeze concentration uses a refrigeration cycle to reduce the temperature of a solution by freezing the solvent. Removing the frozen solvent leaves a concentrated product. Freeze concentration is used in the citrus industry to concentrate juices, in other food industries, and for salt water desalination and sewage treatment. It has numerous advantages over the two other major alternatives—physical separation and vaporization. The benefits of freeze concentration include lower energy consumption, better product quality, greater product recovery, and reduced capital, maintenance, operation, and transportation costs. Indirect freeze concentration is best suited for food processing. A heat-exchange surface keeps the product separate from the refrigerant, and a mechanical device scrapes the surface to prevent ice deposits from forming. Freeze concentration has also been shown to be an efficient volume reduction technology for bleaching plant effluents in paper-pulp mills

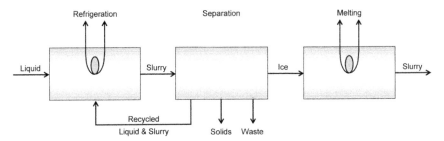

Figure 6-5. Freeze Concentration Process (Source: Heist Engineering Corp., 1987)

and for removing adsorbable organic halogens and or non-process elements from recycled water.

MICROWAVE HEATING AND PROCESSING

Microwaves are in the radio frequency portion of the electromagnetic spectrum between 300 and 300,000 MHz (Megahertz, i.e., million cycles per second). Microwaves are electromagnetic waves similar to radio, TV, light, radar or infrared waves with the primary difference being the frequency of the wave.

Microwaves reside in the radio-frequency portion of the electromagnetic spectrum between 300 and 300,000 MHz. In the U.S., the commercial microwave sources are limited by the Federal Communication Commission to 915 MHz and 2450 MHz. Microwaves are produced by magnetrons which are composed of a rod-shaped cathode surrounded by a cylindrical anode. Electrons flow from the cathode to the anode, creating an electromagnetic field. The field frequency is a function of the dimension and shape of the magnetron. Microwaves are well-suited to heat dielectric-materials that are electrically nonconductive.

A microwave processing system is usually comprised of the power supply and the magnetron; an applicator which directs the microwaves to the product being heated; a materials handling system, and a controller. Microwaves have a higher power density and heat material faster than radio-frequency waves. Radio-frequency lower frequency waves are better suited for heating thicker materials.

The most widespread use of industrial microwave processing is in the food industry for heating, tempering, drying, and precooking. Other non-food applications include rubber vulcanization, welding of plastics, drying textiles, and sintering.

Microwave energy reacts with matter in three basic ways:

1.   It can be reflected with little or no loss of energy, for example, off the sides, top and bottom of the interior of a microwave oven. The microwaves literally bounce around like millions of microscopic rubber balls.

2.   It can be transmitted through materials such as air, glass and some plastics with little or no loss of energy, e.g., through the walls of baking dishes.

3.   It can be absorbed. Food products contain molecules such as water, salts and proteins that exhibit dipolar properties, i.e., they act as if one side or one end had a positive charge while the other side or other end had a negative charge. When these dipolar molecules are exposed to microwave energy, they "vibrate" or rotate which causes internal friction. The internal friction, in turn, produces heat, and the temperature of the food product increases. Since the dipolar molecules are distributed throughout the product, the product heats from within as well as on the exterior surfaces. In contrast, when a food product is heated in a conventional oven, the heat is transferred from the surface to the inside of the product by conduction, or in the case of a fluid material by conduction plus

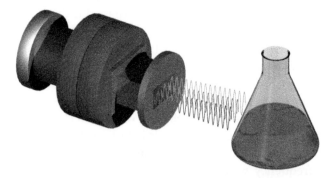

Figure 6-6. Example of the Application of Microwave Assist Technology (MAT)

convection. In either case, this relatively slow rate of heat transfer often results in the surface being overcooked before the center is heated sufficiently.

Since microwaves heat the product from within as well as at the surface, the center heats at approximately the same rate as the exterior, and the total heating times are reduced ten- to twenty-fold.

Applications of Microwave Heating and Processing

Examples of the application of microwave heating and processing include:

Product Drying

The advantages of microwave drying include greater throughput, reduced capital costs, reduced floor space requirements, energy costs reductions of up to 66%, and reduced sanitation costs. The finished products have superior color, flavor and nutrient retention as compared to those processed by conventional drying methods.

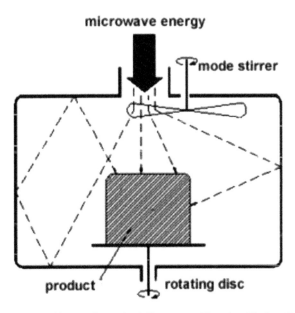

Figure 6-7. Illustration of a Microwave Heating Technology

Puffing/Pre-gelatinization

In food processing, starch-based ingredients such as grains and legumes can be puffed and pre-gelatinized using microwave energy. Since the puffed product is more porous and the starch has been pre-cooked, rehydration is rapid and complete. Foods such as pasta, snacks, tomatoes and fruits, including cranberries, grapes, pineapples, dates, strawberries and bananas, have been dried and puffed by treating them with a combination of microwave energy and vacuum. Retention of the original flavor and color of these unique puffed products is quite good, and consumers have responded favorably to the overall eating quality (EPRI CU 6247).

Precooking

In food processing, precooking meats using microwave energy in combination with conventional heat sources is another example of the benefits that can be obtained with microwave energy is properly utilized. Results of the combination approach, except in cases of very thin product such as bacon where microwaves alone work very well, are usually superior to attempts to cook meats and other products using only microwave energy. Using microwave energy, the same tempering can be achieved in a few minutes. Advantages of microwave tempering include faster response time that allows greater scheduling flexibility, reduced microbiological counts, higher yields since there is less drain loss during tempering, reduced sanitation costs, and better retention of the original flavor and color in the defrosted products.

Ohmic Heating

Ohmic heating is most often used in the food processing industry. In an ohmic heating unit, electrical energy is converted directly to thermal energy within the product due to the electrical resistance of the food product being processed. Since the product is heated from within, there are no problems such as burn-on which can occur at the hot surfaces of conventional heat exchangers. Temperature rise is very rapid, energy utilization is highly efficient, the heater has no moving parts which could damage shear sensitive product, and the units are usually designed for CIP (clean-in-place) sanitation.

Ohmic heating, also called resistance heating, has been used for pasteurization since the 1920s. It initially failed to gain wide acceptance due to the lack of erosion-resistant electrode materials and the limitations of the glass electrode housings to withstand thermal and physical shock. Suitable electrode materials were subsequently developed, and advances in plastic technology have provided high-temperature-resistant electrode housing materials that ensure complete electrical insulation between the product and the metal piping.

High-Energy Electric Pulsed Field

Ultra-high-pressure processing, also called high-pressure processing or HPP, is a method to extend the shelf life of food products without the application of heat or chemical treatment. In this process, food products are exposed to pressures in the range of 15,000 to 90,000 psig for a few minutes. This high-pressure inactivates many types of vegetative microbes at room temperature. Spores and common food enzymes appear to have tolerance to high pressures and are not destroyed.

In theory, this process could be applied to any food product, but in practice, it is limited to those products which do not have components that would be deformed by high pressures. Typically, its application is limited to pumpable acidic materials with little or no particulate material.

Industrial Heat Pumps

Industrial heat pumps use a vapor compression cycle to raise the temperature of a working fluid. This technology is well-suited where large quantities of hot water, low-pressure steam, or water vapor are needed for industrial dehumidification and drying processes at low and moderate temperatures and also for fractional distillation in the petroleum and chemical industries. Other applications include drying of pulp and paper, food processing, wood and lumber, and evaporation and distillation processes. The most prevalent type of heat pumps used in industry is electric closed-cycle compressions heat pumps as shown in Figure 6-8. These heat pumps save energy, have lower emissions and a reduced footprint, and yield better product quality.

The use of heat pumps to recover waste heat in industrial applications represents one of these opportunities. Heat pumps are systems that operate in a cyclic manner. They absorb heat of low temperature from an energy source, apply external energy, and deliver heat to a higher-temperature load, as illustrated in Figure 6-10. The external energy is usually supplied by an electric motor.

Application Example: Open-Cycle Mechanical Vapor Recompression (MVR) Heat Pump in Distillation Column

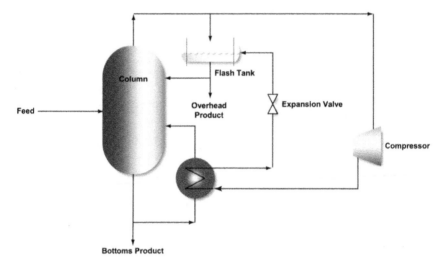

Figure 6-8. Diagram of a Heat Recovery Heat Pump

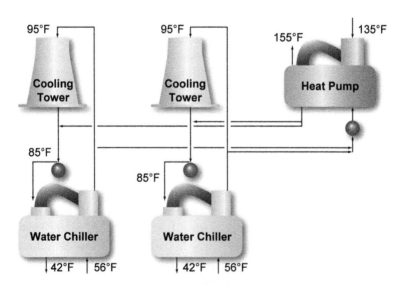

Figure 6-9. Typical Heat Recovery Heat Pump System

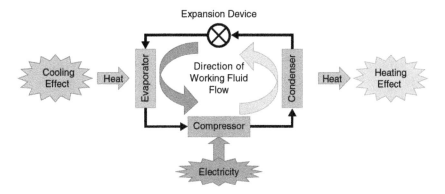

Figure 6-10. Operation of Closed-Cycle Mechanical Heat Pump

There are four common types of industrial heat pumps:

- **Closed-cycle mechanical heat pumps** use mechanical compression of refrigerant. They are used for lumber drying, space heating, and heating water and process liquids (Figure 6-10).

- **Open-cycle mechanical vapor recompression (MVR) heat pumps** use mechanical compression to increase the pressure of waste vapor. They are used in evaporation and distillation processes commonly found in the petroleum, chemical and petrochemical, pulp, and food and beverage industries.

- **Open-cycle thermo compression heat pumps** use high-pressure steam to increase the pressure of waste vapor. They are used in evaporators and flash-steam recovery systems, such as paper dryers.

- **Closed-cycle absorption heat pumps** use a two-component working fluid and the principles of boiling-point elevation and heat of absorption. They can deliver a much higher temperature rise than other heat pumps and have the ability to provide simultaneous cooling and heating. They are typically used in chilling applications.

Heat pumps are extremely beneficial in recovering low-temperature waste heat. It is ordinarily not practical to extract work from waste heat sources in the low-temperature range, and many applications of low-temperature waste heat are limited to using the waste heat for preheating liquids or gases by means of heat exchangers. Under many circumstances, however, heat pumps enable the economic use of low-temperature waste heat in industrial applications requiring higher-temperature heat. Examples of industrial waste heat sources in the low-temperature range include process steam condensate (~130-170°F) and cooling water from various industrial machines, furnaces, internal combustion engines, and hot-processed liquids and solids (~90 450°F) (Global Energy Partners, 2007).

There are numerous industrial applications of heat pumps and a cousin technology—heat recovery chillers. Industrial heat pumps are predominately used by the lumber, petroleum refining, chemical and petrochemical, pulp and paper, and food and dairy industries. Common industrial processes benefiting from heat pumps include drying, evaporation, and distillation. Heat pumps are also employed across most industrial sectors for water heating and space conditioning, similar to how heat pumps are used in residential and commercial buildings. Additionally, the application of absorption heat pumps for chilling is emerging. Heat recovery chillers are less popular. Engineering practice has been to keep cooling and heating process flows separate, thus not examining industrial process flow in a holistic way. As a result, there are tremendous opportunities for use of heat recovery chillers.

The use of closed-cycle heat pumps for drying of leather, foodstuff, and paper is a relatively popular application, but these tend to be small-scale applications. Other closed-cycle applications are generally of a larger scale, but they are typically low energy-intensive applications, such as heating process water or process fluids and heating water for cleaning purposes.

In contrast, evaporation and distillation applications are highly energy-intensive and often involved large-scale systems. Distillation and evaporation lend themselves extremely well to open-cycle mechanical vapor recompression (MVR) heat pumps. In the evaporation process, MVR heat pumps are mainly used for water-based solutions where the vapor produced by the evaporation of liquor is compressed and used to drive the evaporator. The distillation process is a physical separation process that involves the separation of mixtures based on differences in their volatilities. It is an extremely energy-intensive process found in numerous industries, including food and beverage, chemical and petrochemical, oil production, and refineries. For example, distillation

Figure 6-11. Picture of a Heat Recovery Chiller (source: Alabama Power Company)

separation of propylene and propane is extensively used by the chemical industry to produce propylene, a key material in the production of many chemical products.

Because of the many advantages of open-cycle heat pumps, a detailed analysis of the benefits associated with the use of an MVR heat pump in a chemical plant has been conducted. The MVR heat pump is used in a distillation column to separate propylene and displaces steam generated by a natural gas boiler (Pernis, 2007). The primary plant-level benefits associated with this heat recovery heat pump application are summarized in Table 6-3.

DESALINATION

Desalination is the removal of salt from seawater or brackish water to produce pure water for municipal and/or industrial uses. A number of energy-intensive processes have been developed for desalination in-

Table 6-3. Benefits Associated with Use of an MVR Heat Pump

Industry	Chemical/petrochemical (NAICS 325)
End-Use	Separation of propylene and propane in distillation column
Electric Technology	50.2 MW open-cycle MVR heat pump, COP = 8
Electricity Requirement	50,400 MWh per hear (172,000 MMBtu per year)
Displaced Technology	Natural gas boiler for steam generation, 75% efficient
Natural Gas Savings	1,824,000 MMBtu per year
Net On-Site Energy Savings	1,652,000 MMBtu per year (91% savings)

cluding reverse osmosis (RO), distillation, electrodialysis, and mechanical vapor recompression. Electricity consumption can range from 4 to 45 MWh per million gallons (1 to 12 MWh per thousand m3) of water treated, depending on the process. Electricity can also account for up to 50% or more of the annual operating costs for desalination facilities. Desalination will be a necessary alternative for water-short areas and will grow in market size as populations increase and water sources diminish. Availability of low-carbon electricity could facilitate this market growth (EPRI 1014570).

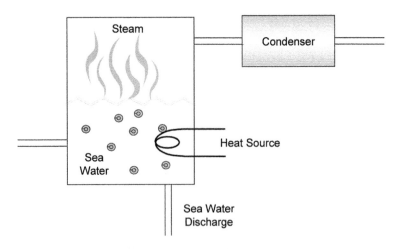

Figure 6-12. Desalinization by Distillation

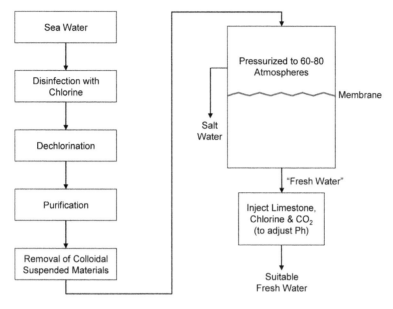

Figure 6-13. Desalinization by Reverse Osmosis

ELECTROLYTICS

Electrolytics are used in the process and materials production industries. In these processes, process reactants are maintained in ionic form in an electrolyte. Voltage is applied to electrodes that are immersed in the electrolyte, separating the ions. Electrolytics are used to produce materials such as chlorine, caustic soda, metals, as well as to deposit finishes.

Electrolytic Separation/Synthesis

Electrolytic separation/synthesis are among the most frequently performed processes in the chemicals industry. Many chemicals and compounds such as chlorine are produced via electrolytic separation. One application of electrolytic synthesis is electrodialysis in which ions are transported through ion permeable membranes from one solution to another under the influence of a potential gradient. This is enabled by the application of voltage between two end electrodes to generate the required electrical field. The membranes used in electrodialysis can selectively transport ions having positive or negative charge and reject

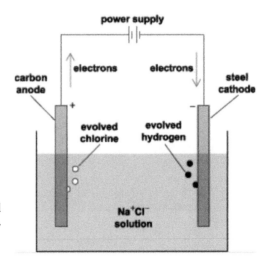

Figure 6-14. Electrolysis Used to Create Chlorine and Hydrogen From Sodium Chloride

ions of the opposite charge. As such, useful concentration, removal, or separation of electrolytes can be achieved. Electrodialysis can reduce electrolyte content in fluids, recover valuable materials from waste streams, split salts, and enable ion substitution.

Electrochemical Synthesis

Electricity demand for chemicals manufacturing is driven both by electrification and by growth in output. For example, the chemicals manufacturing industry can use electricity to produce oxygen for the partial oxidation of fuels.

Chemical producers spend a majority of their capital and operating budgets on the reactors that convert coal or oil into organic chemicals. These city block-sized, three-dimensional circulatory systems contain an array of piping and connected equipment that separates, recycles, and recovers chemicals. This huge system (both in size and cost) can impede a manufacturer and often provides the incentive to develop electro-organic chemical synthesis systems that offer the potential to yield pure specific chemicals.

Despite electroprocessing's relatively high costs, it confers advantages in product quality and process cleanliness that are not possible with thermal techniques. For example, the conventional halogenation route to anisic alcohol leaves trace contaminants with unpleasant odors—a problem if the alcohol is used to impart fragrance. This problem does not arise from use of the electrochemical method. Electrosynthesis of isocyanates and carbonates eliminates the use of highly toxic phosgene that would be needed in thermal processes. This "process cleanliness" is also a major advantage of using electro-organic processing to produce Vitamin C—in this case, manufacturers do not have to use a hypochlorite oxidizing agent, which is troublesome to dispose of.

An electrochemical cell can gasify coal; the anodes release oxides of carbon and the cathode releases hydrogen. Since coal provides about half of the energy for the gasification, use of this electroprocessing method is similar to using plasma arc technology.

NEW APPLICATIONS FOR MECHANICAL ENERGY

Operations accomplished by physical as opposed to thermal or chemical processes typically use less fuel to provide the mechanical energy. Rapidly rising fuel costs make new applications for mechanical processes more attractive. Further, since utility-generated electricity uses low-cost fuels such as coal and uranium, new applications for mechanical processes are energized to an increasing degree with utility-produced electricity.

New mechanical process applications are enormously varied and ubiquitous; however, not all potential applications for mechanical processes are equally attractive. The main objective of adopting these process applications is to reduce the overall cost of production. A particular application, such as using a heat pump in lieu of a gas burner and heat exchanger, may save energy dollars but cost more to purchase and maintain, and thus fail to satisfy justification criteria. Another application, such as the use of membrane separators in lieu of chemical absorption, may reduce each of the production cost components (i.e., capital, labor, and energy) and also provide other benefits such as improving product yield or purity.

Examples of beneficial mechanical processes include installing:

- Vacuum pumps to replace steam ejectors.

- Rotating stages in distillation columns to reduce the number of stages required.

- Freeze crystallization in water processing to reduce energy costs for evaporation.

- Ultra-filtration to improve a wide variety of wastewater clean-up chores.

- Pressure swing absorption and cryogenic processes to recover hydrogen and argon from purge gas in the production of ammonia.

- Physical absorption of carbon dioxide in propylene carbonate to eliminate stripping of potassium carbonate solution ion the production of ammonia.

- Pressure swing absorption to displace wet process technology in the separation of pure hydrogen produced by steam reforming.

ELECTRIFYING IC MOTOR APPLICATIONS

Electric motors are a beneficial new use when used to displace an existing or intended natural gas- or diesel-fired internal combustion engine-powered motor. In order to be competitive with fossil-fueled motors, electric motors need to be applied as effectively as possible.

Many processes, especially those that use pumps and fans, need to vary their output to accommodate changing process needs. This is often done by running the pump or fan at full speed while modulating its output with a partly closed valve or damper. Electronic adjustable-speed drives can reduce this waste, and their use makes electric motors competitive. A new breed of high-efficiency motors are another important advance. These motors are better designed and better made from higher-quality materials than conventional motors are, thereby reducing their magnetic, resistive, and mechanical losses.

Both adjustable-speed drivers and high-efficiency motors are important. However, they account for only half of the total potential electricity savings in U.S. motor systems. The other half comes from 33 other improvements in the choice, maintenance, and sizing of motors, 3 further kinds of controls, and the efficiency with which electricity is supplied to the motor and torque is applied to the machine (Fickett, 1990).

In order to make electric motors more competitive with fossil motors, the design needs to be as robust as possible. The diagram in Figure 6-15 illustrates three versions of applying an electric motor to pumping. The first diagram illustrates a too-often typical motor applied to a pump. In this diagram, the motor runs full bore to engage a standard pump.

The system output is throttled by a valve. In the second diagram, an adjustable-speed driver mechanism is applied, as is an efficient pump and low-friction pipe. The final diagram adds an efficient motor.

Gas Compression Stations for
Natural Gas Transmission and Storage

Gas compression stations use large prime movers, such as gas engines, gas turbines, or electric motors, to drive compressors. Though currently used in only a fraction of gas compression applications, improvements in electric motor technology, variable frequency drives (VFDs), and advanced control algorithms, coupled with low costs for installation and maintenance, have made electric-driven systems more viable despite the higher energy costs for electricity. Indeed, the overall percentage of electric-driven compression horsepower in use for interstate pipelines increased from about 3.5% in 1996 to 5% by the end of 2000. Moreover, the share of electric-driven compression in new installations was 15% in 2000. Lower electricity costs and price stability will increase the market for electric compression further. Electric gas compression stations typically have electric demand of 5 to 50 MW.

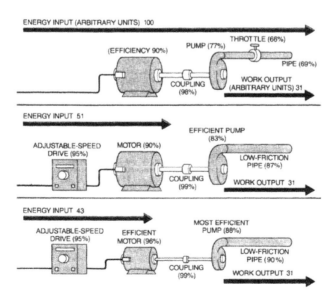

Figure 6-15. Comparison of Pumping Systems
(Source: *Scientific American*)

The U.S. demand for gas is projected to grow between now and 2015 according to several forecasters, including the Gas Technology Institute and the Energy Information Administration. The demand for gas compression must also grow incrementally to meet the greater flow of gas. At the same time, environmental regulations are becoming increasingly stringent, making traditional gas-driven compressor systems more difficult to permit (EPRI 1014570).

Oil Pumping Stations

Large oil pumping stations exist in some areas of the world. These stations may use electricity to power pumps. Individual pumps can be several megawatts in size. An electric oil pumping station represents an electric demand of 25 to 50 MW. Oil is pumped from the ground through pipes to a tank battery. One or more tank batteries may be installed in a single field, each serving a number of individual wells. A typical tank battery contains a separator to separate oil, gas, and water; a fired heater to break water/oil emulsions to promote removal of water from the oil; and tanks for storing the oil until it is shipped as crude oil by truck or, more commonly, by a gathering line connected to storage tanks. From these tanks, the oil is moved through large-diameter, long-distance pipelines—so-called trunk pipelines—to refineries or to other storage terminals. Oil pipelines rely on pumps to initiate and maintain pipeline pressure at the level required to overcome friction and changes in elevation. Pumps are required at the beginning of the line and are spaced along the pipeline to adequately propel the oil along. Pumping stations mostly use crude-oil-fired or natural-gas-fired engines. Some pipelines are replacing these fossil-fired engines with electric engines. Electricity represents a major portion of operating costs at these sites (EPRI 1014570).

Gasoline Pumping Stations

Most gasoline and diesel fuel supplies are delivered to the marketplace by pipelines, from refineries to local distribution centers. Tanker trucks carry gasoline only the last few miles or kilometers of the trip to individual service stations. Major American airports rely almost entirely on pipelines and have dedicated pipelines to deliver jet fuel directly to the airport. These pipelines vary in size from relatively small, ~8-to 12-inch (20 to 30 cm) diameter, lines up to greater than 40 inches in diameter. Pump stations typically range from 1000 hp (0.7 MW) to

5000 hp (3.8 MW) for electrically driven motors and drives per station. Electricity represents a major portion of operating costs at these sites (EPRI 1014570).

MUNICIPAL WASTE AND WATER ELECTRIFICATION

A number of opportunities have been revealed for the beneficial use of electrotechnologies in municipal systems for waste and water treatment. Stricter environmental regulations and the decrease of available space for waste disposal have motivated an ongoing reevaluation of waste management technologies and an emerging interest in innovative waste processing methods.

There are four major classes of municipal waste:

- Municipal solid wastes
- Municipal liquid wastes such as sewage
- Municipal wastes from nonhazardous industrial sources
- Municipal wastes from hazardous industrial sources

Within each of these categories, there are differences in the types and quantities of waste generated, the current and promising disposal or treatment technologies, and the energy usage for these technologies.

Three areas have high practical technical potential for introduction of electrotechnologies: hazardous waste treatment, sewage treatment, and resource recovery. Processes being researched for hazardous waste management include pyrolysis (including plasmas and infrared heaters), electrochemical concentration (including electrodialysis), freeze concentration, and supercritical fluid oxidation. Processes for water treatment include ion exchange, reverse osmosis, and ultra-filtration. For waste recycling, plasma, reverse osmosis, and electrodialysis can be employed in metals recovery and for melting both glass and metal.

Raw Water Pumping Stations

Raw water pumping stations supply water to municipal, industrial, and agricultural applications. These stations use electricity to power large pumping plants. Electricity consumption can range from 1 to 2 MWh per million gallons (0.25 t 0.5 MWh per thousand m3) of water pumped, and electricity is the major cost component for these operations. Many state and federal water facilities serve a critical and increasing supply need (EPRI 1014570).

The treatment of liquid waste is a source of concern due to a number of factors: the difficulty of containing liquid waste, the tightening of environmental regulations, the decrease of available space for waste disposal, and the steady increase in the amount of liquid waste generated.

The treatment of liquid waste is a source of great concern for municipalities and industrial establishments due to the difficulty in handling containment of liquid waste, the tightening of environmental protection regulations, and the decrease of available space for waste disposal, as well as the steady increase in the amount of liquid waste generated. The new environmental regulations emphasize source reduction, recycling and treatment, and impose tighter hazardous waste handling and disposal requirements, resulting in significant increases in disposal costs. The combination of these factors has led to the development and more widespread use of concentration and separation techniques—methods of reducing the total volume of hazardous waste.

The focus of separation and concentration technologies for treatment of industrial wastewaters has been on industries that generate large amounts of wastewater streams, as well as industries under the most regulatory pressure to meet the new stricter industrial pretreatment standards set by federal, state, and local agencies. The industrial markets that met these criteria are food and kindred products, paper and allied products, chemicals and allied products, petroleum refining, and metal finishing which includes printed circuit board manufacturing. Four of these five largest users of water (food, paper, chemicals, and petroleum refining) are in the process industries category and account for over 60% of the industrial water intake in the U.S.

The liquid waste streams produced by the five major industrial markets can be characterized as capital cost, energy usage, area of application, and equipment manufacturers. Information on the manufacturing processes use, the type and amount of resulting waste streams, and current industry practices for processing of these wastes has been gathered. Several existing and emerging separation and concentration electrotechnologies for treatment of industrial liquid wastes are available. Primary examples include:

- Membrane separation (reverse osmosis, ultra-filtration, micro-filtration)

- Heat pump evaporation
- Freeze concentration
- Electroacoustic dewatering
- Biological treatment
- Electrodialysis
- Ion exchange
- Electrolysis

Commercially available separation and concentration electrotechnologies offer an attractive solution to the growing problem of treating industrial wastewaters in the face of stricter environmental regulations. However, these technologies, with the exception of biological treatment, do not have widespread application in the industry today. This is mainly due to lack of reliable information sources on the performance of these systems and the industries' reluctance to use new technology. Providing support for commercial-scale demonstration projects to prove the viability of electrotechnologies and technology transfer activities would positively impact the acceptance of these technologies by industry.

A number of technologies have been identified and evaluated, and five technologies have emerged as the most promising:

1. **Cross-Flow Micro-Filtration Using Ceramic Membranes** is very effective at making oil-water separations and is nearing commercial acceptance by industry. Test programs are necessary to obtain membrane fluxes, separation factors, and run lengths to prove that the membrane will make the separation without plugging.

2. **Supercritical Water Oxidation** operates at high pressure and temperature and requires expensive construction materials, but may represent the best opportunity for essentially complete destruction of hazardous chemicals contained in an aqueous medium. Test programs are needed to demonstrate to end users that the process will run at the severe conditions of pressure and temperature without excessive downtime. Commercialization of this technology will be expensive because of relatively high operating and capital costs of even small-scale units.

3. **Microwave Distillation** could be utilized where distillation separations need to be carried out very fast, such as in the distillation of

pharmaceuticals and other thermally sensitive materials. Specific applications that are cost-effective need top be identified and performance data obtained.

4.   **Electrically Driven Drop Dispersion** is being developed at the Oak Ridge National Laboratory. This technology uses an electric field to break up and disperse water droplets in organic media and could be applied to both liquid-liquid and supercritical extraction. To speed commercialization, a liquid-liquid extraction device capable of processing five to ten gallons per minute needs to be constructed, performance data obtained, and results compared with competing technology.

5.   In the **Electrosorption** process, an electric field is used to control the adsorption characteristics of activated carbon and thus could be used to control the adsorption-desorption cycles for a carbon bed used to track industrial wastewater. Performance data need to be run at flow rates of 10 to 15 gallons per minute to see if this technology is indeed reliable and cost-effective.

Ozone: Antimicrobial Agent

Ozone has a number of applications in industry. It is typically used for water treatment. Ozone is a naturally occurring gaseous material. It is colorless with an odor somewhat similar to that of chlorine. Its odor (detectable by most people at about 0.01 or 0.02 ppm) is often noted near an electric spark or after a lightning storm. Ozone is slightly soluble in water (0.88 volumes per 100 volumes) with its solubility decreasing as the temperature increases. Since ozone is usually used as an aqueous solution, it is more effective at the lower temperatures that allow higher solubility.

Chemically, ozone is the triatomic form of oxygen (O_3). Ozone is unstable and breaks down naturally to oxygen (O_2) and a highly reactive oxygen atom (O). The reactive oxygen atom can act as a disinfectant, i.e., destroying bacteria or act as an oxidant by combining with various chemical compounds including organic compounds. As an oxidizer, ozone is approximately 1.5 times more reactive than chlorine. It is the strong disinfectant and oxidation properties of ozone that make it of value to the food industry. Ozone reacts up to 3,000 times faster than chlorine with organic materials, e.g., bacteria, and produces no harmful decomposition products. The reaction by-product of the ozone molecule is oxygen.

Due to its high reactivity, it is not practical to store ozone; therefore, it is generated on-site. It has a half-life of less than 24 hours in the dry gaseous state and rapidly decomposes when moisture is present, especially at alkaline pHs. Ozone can be produced by a variety of methods including corona discharge, photochemical formation as the result of ultraviolet radiation, electrolysis of water, electrolytic reduction of concentrated sulfuric acid, nuclear radiation using Cobalt 60m and passing moist air over elemental phosphorus.

Commercially, corona discharge is typically used. It consists of passing dry air or dry oxygen through an electric arc between two parallel electrodes—either tubular or flat plate. Generation using dry oxygen is usually the preferred method since it eliminates the potential problems of nitrogen oxides that are generated when air is used.

A typical ozone generation and application system consists of:

- A power supply that controls the electrical energy used to convert the oxygen into ozone.

- Feed gas pre-treatment which passes high-purity oxygen through additional filters to remove any air-borne particulates as well as any trace moisture.

- Ozone generator consisting of two parallel electrodes between which the dry oxygen passes.

- Surplus ozone destruction system which removes and destroys any ozone which is not dissolved or used in the application/distribution system.

- Monitoring and control system.

Gaseous ozone generators of various capacities for on-site production have been marketed for many years in the United States and in Europe. Gaseous ozone is bubbled or defused into the water to be treated. Uniform distribution of fine bubbles is optimal. Ozone monitors are commercially available so treatment levels can be accurately checked.

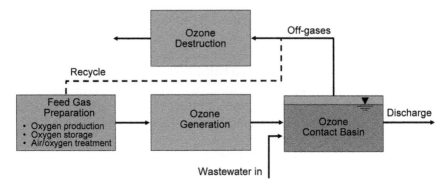

Figure 6-16. Illustration of an Ozonation Process

CONCLUSION

Electrification of the industrial sector is not only beneficial to the environment, but it involves the use of technologies in production which improve product quality, productivity, and profitability.

References

"Electricity and Industrial Productivity," EPRI, Palo Alto, CA: 1984. EM-3640.

"Freeze Concentration: An Energy Efficient Separation Process," EPRI Journal, January/February 1989.

"Roles of Electricity: Electric Steelmaking," EPRI, Palo Alto, CA: 1986. EU-3007-8-86.

"Roles of Electricity: Production of Chemicals," EPRI, Palo Alto, CA: 1986. EU-3015-7-86.

"Industrial Energy Efficient Technology Guide 2007," EPRI, Palo Alto, CA: 2007. 1013998.

"Program on Technology Innovation: Identification of Embedded Applications for New and Emerging Distributed Generation Technologies," EPRI, CA: 2006. 1014570.

"Food Industry 2000: Food Processing Opportunities, Challenges, New Technology Applications," EPRI, Palo Alto, CA: 2000. 1000053.

"Membrane Separation—Food Processing Industry," prepared by Global Energy Partners, LLC for EPRI, August 12, 2009.

"Food Industry Scoping Study," EPRI, Palo Alto, CA: 1990. CU-6755.

"Industrial Microwave Systems," website, http://www.industrialmicrowave.com/.

"Membrane Separation—Food Processing, Industry," EPRI, Palo Alto, CA: 2010. 1020680

"Dielectric Heating—Using Microwave Technology in Non-Metallic Minerals Industry," EPRI, Palo Alto, CA: 2009. 1020503

"Efficient Use of Electricity," A.P. Fickett, C.W. Gellings, A.B. Lovins, Scientific American, Vol. 263, No. 3, pp. 64-74, September 1990.

"Industrial Applications of Freeze Concentration," Heist Engineer Corp., 1987.

"Membrane Separation Technologies," Study #1554, The Freedonia Group, Inc., Cleveland, OH: 2002.

"Industrial Heat Pumps for Waste Heat Recovery," Global Energy Partners, Tech Review, 2007.

"Mechanical Vapour Recompression Case Study," Heat Pump Center, Pernis, the Netherlands, www.heatpumpcentre.org/Publications/case_permis.asp.

Chapter 7

Beneficial Industrial Uses of Electricity: Metals Production

METALS PRODUCTION

The metals production industry category is further divided into the ferrous and nonferrous metals industries, with the former subcategory dealing almost exclusively with the iron and steel industries.

Steel Finishing

Conventional practice for steel casting and finishing operations in integrated steel mills (i.e., mills with facilities for coking coal and smelting iron ore, in contrast to those engaged solely in the remelting of scrap) consists of pouring the steel into large ingots. These ingots are subsequently reheated and, following a series of forming and reheating steps, are made into basic steel products for shipment.

These steel finishing operations represent a substantial part of the overall cost of steelmaking. The costs occur because this process requires several separate operations, and each carries a cost for labor, energy, and capital equipment and contributes to product waste. For example, these operations can produce a ton of raw steel from ore using about 18 MBtu of primary energy. But, about 40 MBtu of energy is expended for each ton of steel product shipped. In addition, only about 70% of raw steel production typically ends up as shipped products—producers must recycle the rest. Many modern mills improve these percentages because they make extensive use of continuous casting techniques.

Continuous Casting

In this process, a funnel feeds molten steel into a water-cooled oscillating copper mold where the steel solidifies and is continuously withdrawn from the bottom of the mold at a rate of about 10 feet per minute. This ribbon of steel is then cut into appropriate lengths (billets) which, in the newest plants, are moved immediately through the rolling mill and emerge shortly thereafter as finished shapes ready for shipment. In this way, 90% or more of the molten steel ends up as finished product. This procedure represents an enormous productivity gain compared with traditional practice.

In principle, producers could deliver all steel to finishing operations via continuous casting regardless of the steel production process. In practice, however, continuous casting operations are most easily and economically introduced in conjunction with new steelmaking facilities. These new facilities can match the capacity and design of the steelmaking equipment with the capacity and layout of the finishing section and with the product line planned for the operation. A small, regional mini-mill fits this specification: it produces a small variety of high-volume simple shapes. In Japan, where more than 70% of the steelmaking capacity came on-line after 1963, more than 80% of steel production was continuously cast by 1982. In that year, the U.S. continuously cast about 25% of its steel production.

Nonferrous Metals

The nonferrous metals subcategory includes most other metals such as aluminum, copper, lead, zinc, and precious metals. The performance of most markets of this industry group mirror general economic trends, fluctuating with economic activity in their major markets. Thus, during periods of economic recovery, growth trends in most of these markets are positive.

Ferrous Metals

Since 1965, when the steel intensity of the U.S. economy began to wane, the steel industry has experienced a decline in production. Among the numerous factors that have contributed to the erosion of the industry are the competitive trade environment and the low growth of major steel-consuming industries. Steel consumption in many major markets including appliances, oil field equipment, machine tools, nonresidential construction, and the automotive manufacturing industries has either declined or remained stagnant. In addition, competition from substitute materials (such as concrete in the construction industry and aluminum and plastic in the automotive industry) have reduced the steel intensity in major steel markets. As a result, the record growth in these markets has not caused corresponding growth in the steel industry.

Investment in technologies such as electric furnaces, continuous casting, and electrogalvanizing are expected to increase productivity and product quality. For example, investment in electrogalvanized capacity has contributed to the steel industry's competitive position.

Aluminum Production

Aluminum and copper producers comprise the two largest energy consumers in this subgroup and both are steadily electrifying their processes. Because the industry now reduces alumina electrolytically, aluminum production is already largely electrified. In addition, processors now use electricity to replace fuels in the remelting of ingot and scrap. Processors prefer electric melting with both electric resistance heating elements and induction furnaces over combustion processes. Electric melting efficiently uses energy and reduces scrap losses to oxidation (up to 50% of the charged metal) because this process limits exposure to combustion gases.

Copper Production

The copper industry can reduce the energy used in ore smelting by 40% when using an oxygen enrichment process; this process doubles electricity use. Electric smelting has replaced fuel smelting altogether in some installations. Even without electric processing, electricity will become increasingly important for copper production because U.S. ores are low grade and require either additional crushing and grinding or hydrometallurgical processing (i.e., acid leaching followed by electrolysis of the copper solution).

Arc Furnaces

Electric arc furnaces have been in use in the U.S. for over 100 years. They are primarily used for the production of common carbon and low-alloy steels and to melt iron and steel for casting operations.

There are two popular types of arc furnaces: direct and indirect. With direct arc furnaces, a charge of metal is placed into the furnace, the furnace is sealed, and then the arc is struck. These furnaces melt steel or scrap iron by direct contact with an electric arc between an electrode and the metal charge. Direct arc furnaces range from under 10 tons (used in foundries to melt iron and steel for castings) to more than 400

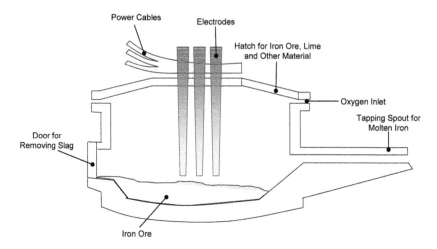

Figure 7-1. Illustration of an Electric Arc Furnace

tons (for industrial-scale steelmaking from scrap steel).

With indirect arc furnaces, an arc is drawn between two carbon electrodes placed above the metal charge, and heat radiates from the arc to the metal. These furnaces have a horizontal barrel-shaped steel shell that is lined with refractory materials. The shell rotates and reverses to heat the refractory evenly. Indirect arc furnaces are used for applications such as producing copper alloys, and they tend to be much smaller than direct arc furnaces.

A third type of arc furnace is less popular—it is the submerged arc furnace. In submerged arc furnaces, electrodes are positioned deep in the furnace, and the reaction takes place at the tip of these electrodes. These arc furnaces are used to produce metals and materials such as silicon alloys, ferromanganese, calcium carbide, and ferronickel. Ore materials are mixed with a reducing agent such as carbon outside the furnace, which is then added periodically to the furnace. The reduction reaction occurs continuously inside the furnace. The furnace is tapped at regular intervals.

Direct arc furnaces used for steelmaking (also called mini-mills) make new steel from scrap iron and steel, while conventional coal-fired basic oxygen furnaces use fresh iron ore. In terms of capital costs, direct arc furnaces are less expensive (in terms of $/ton of steel capacity) than basic oxygen furnaces. One ton of steel in an electric arc furnace requires around 400 to 500 kilowatt-hours per short ton, which is about one-third to one-tenth the energy required by basic oxygen or integrated blast furnaces. Electric arc furnaces used for steelmaking are usually used where there is an inexpensive supply of electric power and a good supply of scrap iron and steel.

Direct arc furnaces used in foundries typically produce iron for castings. These units tend to be smaller (under 25 tons) and also use scrap steel and scrap iron.

The Mini-Mill

The switchover from open hearth to basic oxygen processes for steel production created an opportunity for new steel producers to enter the competition. Producers can build small mini-mills with a relatively modest investment. These mills take advantage of a greater availability of local scrap and, because of their relatively small size, can be located virtually anywhere, thus avoiding the cost problems associated with escalating freight rates.

Electro Slag Remelting

Electro slag remelting (ESR) or electro flux remelting was invented in the 1930s for remelting and refining steels and special alloys. ESR uses highly reactive slag to reduce the amount of sulfide and other in-

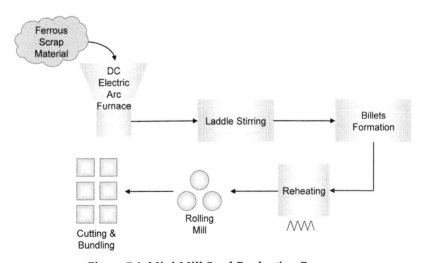

Figure 7-2. Mini-Mill Steel Production Process

clusions present in biometal alloys. ESR is an alternative to vacuum arc remelting that is prevalent in U.S. industry. ESR can produce a wide range of small-weight ingots of tool steels and super alloys, and heavy forging ingots up to raw ingot weights of 165 tons.

In the ESR process, a consumable electrode is dipped into a pool of slag. It is surrounded by a water-cooled mold. Current passes through the slag, between the electrode and the ingot being formed, and super-heats the slag so that drops of metal are melted. These drops travel to the bottom of the water-cooled mold where they solidify. The new ingot of refined material is homogeneous, directionally solidified, and free from central unsoundness. ESR furnaces can be used for remelting of round, square, and rectangular ingots.

Modern plant design, coaxial current feeding and computer control, and regulation have achieved fully automatic remelting processes. With these designs, ESR offers very high, consistent, and predictable product quality. Controlled solidification improves soundness and structural integrity. Ingot surface quality is improved by the formation of a solidified thin slag skin between the ingot and mold wall. ESR has become the preferred method for producing high-performance super alloys.

Induction Heating and Melting

Induction heating and melting was first used in channel-type induction furnaces for metals melting operations. These furnaces evolved to coreless induction furnaces for melting, superheating, and holding. The technology is also applied to harden metal engine parts. More recently, there is increased use of induction heating technology in the ferrous and nonferrous metals industries.

In an induction heating or melting system, a copper coil that surrounds the part to be heated or melted is energized. When a metal part is placed within the coil, circulating eddy currents are induced within it. These currents flow against the electrical resistivity of the metal, generating precise and localized heat. When the part begins to melt, electromagnetic forces agitate it. Mixing and melting rates are controlled by varying the frequency and power of the current in the coil. The control system can moderate the temperature.

Coreless furnaces use a refractory crucible surrounded by a water-cooled AC current coil. They are used primarily for remelting in foundry operations and for vacuum refining of specialty metals.

Channel furnaces have a coil wound on a core. A secondary coil is in the furnace interior, surrounded by a molten metal loop. Channel furnaces are usually holding furnaces for nonferrous metals melting combined with a fuel-fired cupola, arc, or coreless induction furnace, although they are also used for melting as well.

The efficiency of induction heating systems varies by specific application and depends on the characteristics of the work piece, coil design, the types and capacity of the power supply, and the degree of temperature change required for the application. Induction heating only works with conductive materials, usually just metals. Induction heating can be used to heat plastics and other nonconductive materials by heating a conductive metal susceptor that transfers the heat. Induction heating can also be used to heat liquids in vessels and pipelines and is, therefore, often used by the petrochemical industry.

When heating conductive materials, approximately 80% of the heating effect occurs on the surface of the part. Small or thin parts generally heat more quickly than large or thick parts, especially if the larger parts need to be heated all the way through. Induction heating has no contact between the material being heated and the heat source, which is important for some operations. This facilitates automation of manufacturing processes. Once an induction system is calibrated for a part, work pieces can be loaded and unloaded automatically. Other uses of induction heating include heat treating, curing of coatings, and drying.

Magnetic materials are the easiest to heat with induction technology. Magnetics materials also produce heat through the hysteresis effect.

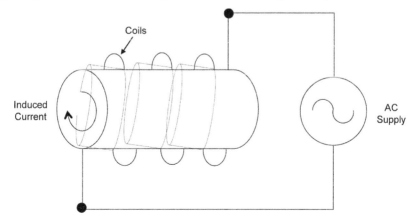

Figure 7-3. Induction Heating for a Tubular Conductor

Induction heating is mainly used in the refining and remelting of metals such as aluminum, copper, brass, bronze, iron, steel, and zinc. Induction systems are also used in applications where only a small selected part of a work piece needs to be heated. Because induction systems are clean and release no emissions, a part can be hardened on an assembly line without having to use a separate heat treating operation.

Laser Heating

Light Amplification by Stimulated Emission of Radiation or *lasers* are produced by the imposition of a voltage field; an electron may jump from its normal energy state to an "excited" state of higher energy. When the voltage field is removed, the electron will decay to its stable state, emitting a packet of light, or photon. If a laser medium is repeatedly excited, a large fraction of the atoms in the materials can be placed in the excited state simultaneously. As some of the electrons in the medium change their normal states, the photons they emit stimulate the transition of other electrons, and a cascade effect occurs, generating an intense light pulse.

Lasers produce high-energy light by passing electricity through a lasing medium (gases or solids). All of a laser's light is of the same wavelength and in the same phase, yielding a very high energy density. Gas lasing mediums are mixtures of carbon dioxide, helium, and nitrogen. Solid-state lasing mediums use materials like yttrium-aluminum-garnet crystals. Both gas and solid-state lasers are commonly found in industrial manufacturing processes.

Laser light has characteristics which make it valuable in industrial applications. The high power density and precise controllability of lasers open a broad variety of opportunities for their use in materials processing. These include applications in cutting, welding, surface treatment, and etching of both metallic and non-metallic materials (Cheremisinoff, 1996).

Laser heating systems have grown from small laboratory lasers to powerful tools found in many industrial sectors. Lasers have high energy density and can be applied in a wide range of manufacturing processes. Lasers are well-suited for operations such as cutting, welding, surface heating, melting, roughening, and cleaning. Industrial-scale units are widely used in highly automated workstations for surface hardening, material removal, and welding operations.

The most common heating application using lasers is for surface hardening. A laser beam is focused on a work piece area, causing the surface to be heated rapidly. For surface hardening, the laser transmits energy to a material's surface and transforms the metal to create a hardened layer. They are the most efficient technology which can be used to harden a specific area instead of an entire part.

Lasers can be computer-controlled, making laser heating well-suited for applications where selective areas within a given work piece are subject to high stress such as crankshafts, gears, and high-wear areas in engine components. Most common steels, stainless steels, and cast irons can be hardened by laser heating.

DIELECTRIC HEATING

Dielectric heating is also accomplished with the application of electromagnetic fields. The material to be heated is placed between two electrodes that are connected to a high-frequency generator. The electromagnetic fields excite the molecular makeup of the material, thereby generating heat within the material. Dielectric systems can be divided into two types: radio frequency (RF) and microwave. RF systems operate in the 1 to 100 MHz range, and microwave systems operate in the 100 MHz to 10,000 MHz range. RF systems are less expensive and are capable of larger penetration depths because of their lower frequencies and longer wavelengths than microwave systems, but they are not as well suited for materials or products with irregular shapes. Both types of dielectric processes are efficient alternatives to fossil-fueled processes

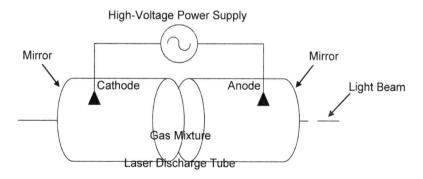

Figure 7-4. Illustration of a Laser

for applications in which the surface to volume ratio is small. This is because of the ability of dielectric heating to generate heat within and throughout the material, while fossil-fueled process heating technologies rely on conductive, radiative, and convective heat transfer to bring heat from the outside in.

Microwave processing is the generation of heat in materials of low electrical conductivity by an applied high-frequency electric field. For a substance to be microwaveable, it must possess an asymmetric molecular structure. The molecules of such substances (e.g., water molecules) form electric dipoles. The electric dipoles try to align with the orientation of the electric field. This orientation polarization mechanism is responsible for generation of energy in the substance. As alternating current is applied to the magnetron, the dipoles change orientation rapidly generating heat.

Industrial microwave heating applications typically use 2,450 MHz systems. However, more expensive 915 MHz systems are becoming increasingly common because their total system efficiencies and penetration depths are higher relative to those of the 2,450 MHz systems.

The National Research Council (www.nationalacademies.org/nrc/) has identified five characteristics unique to microwave processing. They include penetrating radiation, controllable electric-field distribution, rapid heating, selective heating, and self-limiting reactions. These characteristics are beneficial in many cases, but can be problematic in other instances. For example, microwave's penetrating radiation enables uniform heating of large sections but can result in hot spots in heterogeneous materials.

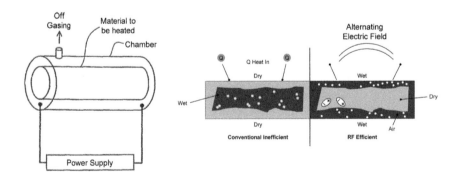

Figure 7-5. Illustrations of Dielectric Heating

The market penetration of microwave processing in industry is relatively limited. Currently, most industrial microwave applications involve heating and drying of food, paper, and textile products. Two of the most promising applications for microwave technology include the sintering of ceramics and the processing of all-water emulsions. Though it can be used across a wide array of ceramics, microwave sintering is currently limited to niche applications in the ceramics industry. However, microwave-assisted firing of numerous types of ceramic materials, including alumina, zirconia, silicon nitride, and silicon carbide, is emerging. In addition, the use of microwaves to promote the dissociation of the only component from water in wastewater or oil emulsions has wide applicability in numerous industries including the petroleum, meat processing, and steel industries. Typically, a conventional all-water separator is used to separate the oil from the water in the oily wastewater. This method is time-consuming because it relies on gravity, and the separation efficiency is poor. Limited, bench-scale research by EPRI and the American Iron and Steel Institute has proven that by pre-treating the waste with microwaves, the oil-water separator can work more effectively (EPRI 1020503).

Other promising microwave applications are emerging, such as in-line processing of pharmaceuticals and enhancing chemical reactions and synthesis. However, many of these applications still remain in the research and development stages.

Table 7-1 lists existing and emerging industrial applications of dielectric heating. Use of dielectric heating in industry is currently limited. However, many applications of dielectric technologies with the potential to yield substantial benefits are emerging. For example, the use of microwave-assisted technology for the firing and calcining of ceramics demonstrates energy savings of 40 to 60% or more relative to conventional gas-fired kilns; moreover, laboratory and pilot-scale tests have demonstrated tremendous potential for dramatic reductions in processing times (EPRI 1020503).

Microwave-enhanced processing is also beneficial to the plastics, chemicals, and pharmaceuticals industries. Indeed, many experts believe microwave-enhanced organic synthesis has a particularly promising future in the production of chemicals and pharmaceuticals. Unfortunately, there are very few publicly available examples of microwave installations in the chemicals and petrochemicals industries. Furthermore, those installations are typically protected by confidentiality agreements

Table 7-1. Examples of Existing and Emerging Applications of Dielectric Heating in Industry (Source: EPRI 1020503)

Industry	Drying	Heating	Separation/Waste Reduction	Enhancing Chemical Reactions
Biodiesel			Optimization of biodiesel production	
Chemicals	Drying of fine chemicals Drying of peroxide/ explosive materials	Heating and liquefying of corrosive viscous products Preheating of resins	Separation of oil-water emulsions	Enhancing chemical processing and reactions (e.g., catalysis, pyrolysis) Reactive treatment of heat sensitive polymers (e.g., resins) Oligomerization of methane Processing of VOCs
Food	Drying of food (e.g., nuts) Drying and puffing of tobacco Drying and roasting of coffee beams Anti-fungal treatment	Heating, coagulation, and pasteurization of food (e.g., sausages, krill, fruit pulp. hazel nut paste, cheese products) Tempering of frozen food (e.g., packed sliced bread, spices/ herbs) Anti-fungal treatment		
Iron	Drying of iron ore			
Lumber	Drying of wood and laminated wood Drying of wood-fiber boards	Preheating of wood and laminated wood		

Table 7-1 (*Cont'd*). Examples of Existing and Emerging Applications of Dielectric Heating in Industry (Source: EPRI 1020503)

Industry	Drying	Heating	Separation/Waste Reduction	Enhancing Chemical Reactions
Manufacturing	Drying of paint Heating and drying of ceramics Drying of ceramic catalyzers			Sintering of ceramics, barium titanate, tungsten carbide Manufacture of catalysts
Many industrial sectors	Drying of filters		Wastewater treatment Heating/combustion of carbon particles in waste gas Treatment of oil/gas production sludge Waste reduction (e.g., discarded rubber tires) Sterilization of critical waste	
Petroleum			Separation of oil-water emulsions	
Pharmaceuticals	Vacuum drying of tablets Drying of filters used for separation	Pasteurization of pharmaceutical products Heating and liquefying of high-viscosity raw materials used in the pharmaceutical and cosmetic industries Tempering of cosmetic textures Ultra-fast heating of sera		Enhancing chemical processing and reactions for the manufacture of pharmaceuticals

Table 7-1 (*Cont'd*). Examples of Existing and Emerging Applications of Dielectric Heating in Industry (Source: EPRI 1020503)

Industry	Drying	Heating	Separation/Waste Reduction	Enhancing Chemical Reactions
Plastics	Drying of plastic raw materials and granulates Drying of film sheet materials and film emulsion Drying layers of varnish Drying casting compounds	Heating of laminated sheets and boards Preheating of plastic profiles Heating and melting of infusion systems, catheters, etc. Melting of optical fibers Heating/curing of epoxy Heating of heat-shrinkable molded parts and tubes		Polymerization of fiber glass reinforced profiles
Paper	Drying and preheating of paper web for printing Drying and preheating of paper sheets Drying of adhesive coatings on fast paper webs	Heating of paper web to remove solvents Heating of fleece webs Preheating leather-fiber webs		
Textile	Drying of fast threads Drying and heating of textile webs	Heating of cotton reels		

preventing microwave equipment manufacturers from discussing the specifics of the applications, such as energy savings.

Plasma Processing

Plasma processing can be used for welding and cutting operations and, more recently, for metals heating and melting applications.

A plasma is created when a gas is ionized. It can be created by exposing a gas to a high-intensity electric arc or by rapidly changing electromagnetic fields generated by induction, capacitive, or microwave generators. These processes can raise the gas's temperature to as much as 20,000°F, freeing electrons from their atoms.

There are two types of plasma devices: transferred arc and non-transferred arc. In transferred arc processing, an arc forms between the plasma torch and the material. The plasma torch acts as a cathode, the material as the anode, and an inert gas passing through the arc is the plasma. These systems are used for metals heating and melting. With non-transferred arc processing, both the anode and the cathode are in the torch. The torch heats a plasma gas creating extremely high temperatures that can provide heat for chemical reactions and other processes.

Applications include melting of scrap metals and remelting. Plasma processing is common in the titanium industry and for melting high-alloy steels, tungsten, and zirconium. Plasma processing can also

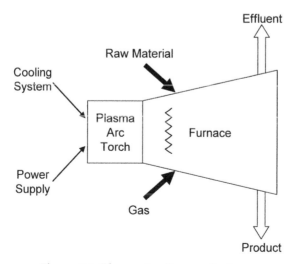

Figure 7-6. Plasma Arc Process System

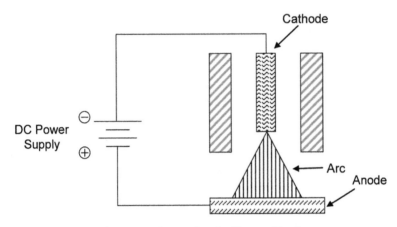

Figure 7-7. Example of a Plasma Torch

be used in the reduction process for sponge iron and smelting reduction of iron ore and scrap steel.

Radio-frequency Heating

The radio-frequency (RF) portion of the electromagnetic spectrum is between 2 and 100 MHz. Radio-frequency waves can only be used to heat materials that are electrically nonconductive (dielectrics).

RF waves are produced by controlled frequency oscillators with a power amplified (called "50-ohm" or "fixed impedance" RF generators) or a power oscillator in which the load to be heated is part of the resonant circuit ("free-running oscillators). Fifty-ohm generators are used exclusively in industrial heating applications in Europe, Japan, and in the semiconductor industry.

A RF processing system is usually composed of an RF generator; an impedance matching network (for 50-ohm generators); electrodes that expose the RF electric field to the product; a materials handling system, and a control system.

RF heaters are similar to microwave ovens where products (food, biscuits, paper, textiles, etc.) passing through the over undergo volumetric heating by virtue of being exposed to an RF energy source. In most products, RF heating is targeted toward water molecules which are polar. (Other compounds can also be targeted.) As the electrode plates are alternatively charged with positive and negative polarity, the electrodes tend to attract one pole of the water molecule and then the

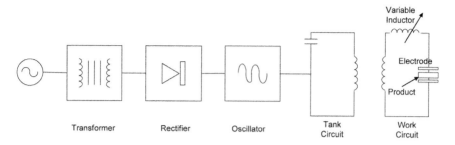

Figure 7-8. Simplified Schematic of a Simple Radio Frequency (RF) Heating System

other. This causes the molecules to move rapidly from one side to the other, causing the molecules to "rub" against one another. This causes frictional heating.

Radio-frequency processing is generally more productive, more energy efficient, and requires less labor and space than convection ovens. Hybrid systems have both radio-frequency processing and a convection oven.

Microwave systems have higher power density than radio-frequency waves and usually heat material faster, while radio-frequency processing's lower frequency waves are better suited for thicker materials.

RESISTANCE HEATING AND MELTING

Resistance heating is the simplest and oldest electric-based method of heating and melting. Resistance heating applications can be precisely controlled and automated, have low maintenance, and no emissions from combustion. Resistance heating can be found in both high- and low-temperature applications ranging from melting metals to heating food products. In many cases, resistance heating is chosen because of its simplicity and efficiency, which can approximate 100%. There are two basic types of resistance heating: direct and indirect.

With *direct resistance heating*, electricity is passed directly through the material to be heated, heating it directly due to the Joule effect. When electricity flows through a substance, the rate of evolution of heat in watts equals the resistance of the substance in ohms times the square of the current in amperes. Typically, the object is clamped to electrodes in the walls of a furnace and charged with electric current. Electric resistance within the material generates heat which heats or melts the metal. The temperature is controlled by adjusting the current.

The material to be heated must conduct a portion of the electric current. The lower the conductivity, the greater the resistance and heat generated. For steel, with its low conductivity, resistance heating is quite efficient. Direct resistance heating is used for applications like heat treating, forging, extruding, wire making, seam welding, and glass heating. Direct resistance heating is also used to prepare steel pieces prior to forging, rolling, or drawing applications. Direct resistance furnaces are also used for holding molten metal.

Indirect resistance heating transfers heat from a resistance element to the material by radiation, convection, or conduction. These elements are made of a very high resistance material such as graphite or silicon carbide. Heating is done in a furnace lined with ceramic, brick, and fiber batting, and furnace interiors can be air, inert gas, or a vacuum.

Indirect resistance heating is also done in encased heaters, where the resistive element is enclosed. The heater is placed in a tank filled with a liquid that is highly viscous or close to a solid. Several types of resistance heating equipment are used including strip heaters, cartridge heaters, and tubular heaters.

Resistance heaters that use convection as the primary heat transfer method are primarily used for applications below 1250°F. Those that use radiation are used in applications that require higher temperatures, sometimes in vacuum furnaces. Indirect resistance furnaces come in a

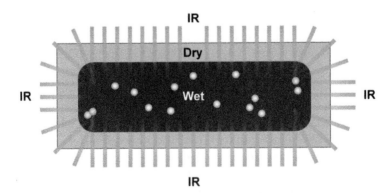

Figure 7-9. Illustration of Indirect Resistance Heating by Infrared (IR)

variety of materials and configurations ranging from counter-top models to ones as big as a semi-truck.

Indirect resistance heaters are used for heating water, heating paraffin and acids, sintering, ceramics, heat pressing fabrics, brazing and preheating metal for forging, stress relieving, and sintering. Element configurations include immersion heaters, circulation heaters, and band heaters. Many hybrid applications also exist.

Electrolytic Reduction

Electrolytic reduction processes typically employ high temperatures using molten salt electrolyte to extract metals from ores. In electrolytic reduction, the density of the electric current can be precisely controlled, leading to high product purity/quality. Aluminum, sodium and magnesium are the dominant products produced by high-temperature electrolytic reduction. Zinc, copper, and manganese are produced in low-temperature electrolytic cells. Copper is also made using low-temperature electrolytic reduction processes. Chromium and lead are also produced through electrolysis. Electrolytic reduction cleaning is one of the most efficient and effective methods of conserving metal artifacts or components.

CONCLUSION

There are many opportunities for electrification in the metals production industries.

References
"Electrotechnology—Industrial and Environmental Applications," N.P. Cheremisinoff, Noyes Publications, 1996.

"Electric Process Heating—Technologies/Equipment/Applications," M. Orfeuil, Battle Press, 1987.

"Dielectric Heating—Using Microwave Technology in Non-Metallic Minerals Industry," EPRI, Palo Alto, CA: 2009. 1020503

"Roles of Electricity: Electric Steelmaking," EPRI, Palo Alto, CA: 1986. EU-3007-8-86.

"Roles of Electricity: Production of Chemicals," EPRI, Palo Alto, CA: 1986. EU-3015-7-86.

Chapter 8

Beneficial Industrial Uses of Electricity: Materials Fabrication

MATERIALS FABRICATION

The metals fabrication category consists mainly of those industrial markets that produce machinery and equipment for other manufacturing industries. As such, the performance of industries in this category is intertwined with the fortunes of all other industrial markets. Also, one materials fabrication group can cover a broad spectrum of industries. The machinery market, for example, provides industrial machinery ranging from metalworking equipment to computer equipment.

This category as a whole is expected to benefit from the continued reliance by user industries such as the automotive, aerospace, and other fabricating industries, on automation. A major factor of concern in this category is the trade issue. Affected by the same trends that have bolstered imports in other markets, many segments of this group have been forced to become more competitive. Industries have attempted to reduce manufacturing costs and increase productivity, lower labor costs (e.g., the employment decline in the metalworking industries), and implement structural changes by increasing participation in joint ventures and in manufacturing and licensing arrangements. In the paper machinery industry, these structural changes have extended beyond U.S. borders to include a reduction in the number of foreign subsidiaries owned by U.S. firms.

Nonmetals Fabrication

This category consists mainly of the durable consumer goods manufacturing market and construction materials such as lumber and wood products, rubber products, stone, clay, and glass products. Industry performance in this group centers on indicators relating to consumer spending and confidence and the health of the construction industry.

These industries are sensitive to fluctuations in their end-use markets. For example, economic upturns in new housing starts as well as the improved levels of existing home sales strengthen the construction market. The ripple effect from such an upsurge causes a rise in the demand for products used in both commercial and residential construction.

Sustained recovery in the U.S. economy and the resultant beneficial impact on consumer spending and confidence helped to maintain strong demand for many of the products in this category. Lumber and wood products, printing and publishing, furniture and fixtures, rubber and miscellaneous products have experienced a recent market expansion. The suppression of growth in individual segments of this industry group, such as textiles, apparel, glass containers, and leather products, arose because of either one of two competitive forces: (1) the proliferation of imports of their products, and (2) the substitution of other products.

Cement Production

Increasing electrification of the cement production process presents some technical difficulties. Producers can replaced gas with coal in the process without degrading cement (and some have begun to make this substitution). However, this switch to coal produces an associated (but modest) increase in electricity use, because conveying and pulverizing the coal requires electricity as well as do electrostatic precipitators that must clean up flue gases. Cement producers have begun using oxygen to enrich the combustion air, since it reduces total energy use by 15%. If the oxygen is generated at the cement production site, the extra electric energy needed is about 10% of the fuel saved.

Plasma arc processing can produce a wide range of hydraulic cements and offers a promising use of electricity in this industry. Processors can incorporate plasma arc reactors into existing conventional cement kilns to provide additional thermal energy. Such a use is analogous to augmenting the output of fuel-fired glass furnaces with electric heat.

Electric Infrared Heating

The implementation of industrial electric infrared (IR) heating systems dates back to the mid-1930s when Henry Ford first used IR systems to cure paint on auto bodies. With the advent of new infrared-specific

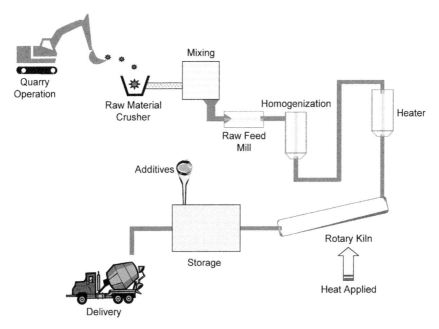

Figure 8-1. Cement Making Process Illustration

coatings, improved emitter designs, and better computer controls, electric IR is now used in many manufacturing applications.

Electric IR heating systems are comprised of an emitter, a reflector system, and controls. Most electric IR applications also come with material handling and ventilation systems. IR is the part of the electromagnetic spectrum between visible light and radio waves ranging from 0.8 to 1000 microns. IR energy can be transmitted, absorbed, and reflected. IR is usually used in applications where the object being heated is in line-of-sight of the IR heater or reflector.

Because IR systems can dry or cure a product in seconds, very accurate control is needed. Many varieties of emitters are available including long-tube-type panel heaters, ceramic bodies with embedded coils, metal coils, ribbons, foils, and fiber heaters. These variations make electric IR a flexible technology with uses in many manufacturing applications.

IR energy is created by conducting electric current through the emitter or filament, and IR systems are classified by wavelength: short, medium, and long. Each class of wavelength has its own heat transfer and absorption qualities.

Short-wavelength emitters often resemble tube-type fluorescent bulbs and are filled with an inert gas such as argon to prevent oxidation of the filament. Operating temperatures of these emitters are around 3500°F and heat-up times are less than 10 seconds. Short-wave systems are often used for spot heating or in supplemental ovens. Medium-wavelength emitters come either as a helically wound coil encased in a long, unsealed quartz glass tube or a resistance coil surrounded by magnesium oxide that is encased in a metal tube. Long-wavelength emitters consist of wires embedded in ceramic panels. Typical applications for medium- and long-wavelength systems are drying and heating.

Electric IR is ideal for situations where a fairly flat product is being heated, dried, or cured. IR primarily heats the surface of a work piece and is not well-suited for products that need to be heated deep beneath the product surface. Products with complex surfaces require a hybrid system with a convection oven or a material handling system that can rotate parts. The work piece or its coating must also have a reasonable absorption in the infrared part of the spectrum. Special paints, adhesives, and other coatings are "tuned" for infrared drying. Common industrial applications of IR include adhesive drying, ink curing, and powder coating curing.

Electric infrared heating is used in many manufacturing sectors for heating, drying, curing, thermal-bonding, sintering, and sterilizing applications. Sometimes fuel-fired heaters are used in conjunction with electric IR in hybrid systems. Ultraviolet (UV) curing is used for curing inks, coatings, adhesives, liquids, and powdered coatings. UV systems require less energy and have lower volatile organic compound (VOC) emissions than IR or convection ovens, but can only be used with certain coatings for niche applications.

Electric IR systems are often deployed with conventional direct-fired process heaters like convection ovens. The IR system begins the drying process, which is finished in a conventional oven. An example is automobile production where IR is used to rapidly set the paint on the body, and then the finish is set in a convection oven. Hybrid systems have the potential to increase throughput by increased line speed.

Electron Beam Heating

With electron beam (EB) heating, the kinetic energy of a stream of electrons is converted to heat when the electrons impinge on the surface

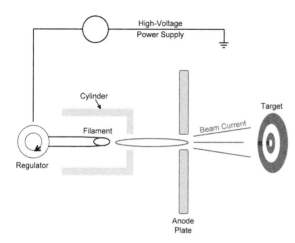

Figure 8-2. Electron Beam Illustration (the filament acts as a cathode and with a voltage applied, the electrons accelerate toward the anode)

of material. Electron beam heating is primarily used in the transportation industry for local surface hardening of high-wear components. EB heating is also used for curing applications such as film lamination and wood finishing.

In electron beam heating, metals are heated to intense temperatures. When used for curing, a liquid is chemically transformed to a solid on the work surface by a stream of directed electrons. EB processing can be done under vacuum or partial vacuum conditions. Vacuum conditions remove gaseous molecules from the air between the electron gun and the work piece, allowing for a tighter beam, which yields less scattering. EB heating in a high vacuum is a non-contaminating melting technique used to produce materials ranging from refractory metal alloys to metallic coatings on plastic jewelry. Electron beam processing enables formation of super-pure materials and can impart unique properties.

EB systems can be computer controlled and can move from point to point easily and rapidly They can be easily pulsed, and they can be used for selective surface hardening. These systems are environmentally because they don't use solvents.

Electron beam curing is generally used to cure thick, heavily pigmented coatings. EB curing systems have a large footprint and can require skilled labor to operate them. However, because they can greatly reduce curing times, they can significantly improve productivity levels.

Electrolytic Deposition/Removal

Electrolytic deposition/removal processes are used to deposit finishes on metal or remove metal from a metallic work piece. Electrogalvanization uses an electrolytic cell and a zinc salt solution to form a protective ring coating on steel. Electrofinishing uses electrolytic methods to produce finishes on manufactured products. Electrochemical machining passes current through an electrolyte to a conductive work piece, dissolving material by electrochemical reaction. This is useful for removing rust from tools or artifacts without altering the surface of the metal.

ELECTRON BEAM MACHINING

Electron beam machining (EBM) uses a high-energy beam of electrons for drilling and cutting metals, non-metals, ceramics, and composite materials. EBM is very similar to laser-beam machining. Electron beam machining requires a vacuum chamber so as to eliminate molecules from altering the course of the beam.

In EMB, a stream of electrons is accelerated and forced through a valve. After passing through the valve, the beam is focused onto the surface of the work piece by a series of electromagnetic lenses and coils. The electrons cause the affected section of the work piece to heat up and vaporize.

ELECTRIC DISCHARGE MACHINING

Electric discharge machining (EDM) is a machining method primarily used for hard metals that are impossible to machine using traditional techniques. EDM is particularly well-suited for cutting intricate contours or delicate cavities. Metals that are commonly machined with EDM include hastalloy, hardened tool steel, titanium, carbide, Inconel, and kovar. One critical limitation of EDM is that it only works with materials that are electrically conductive.

EDM removes metal by producing a rapid series of repetitive electrical discharges that melt away small particles. Electrical discharges are passed between an electrode and the work piece. The material that is removed from the work piece is flushed away with a continuously flowing fluid. The repetitive discharges create a set of successively deeper cuts in the work piece until the final shape is produced.

There are two EDM methods: ram EDM and wire EDM. The primary difference between the two involves the electrode. In ram EDM application, a graphite electrode is machined with traditional tools. The electrode is connected to the power source, attached to a ram, and slowly fed into the work piece. The machining operation is usually performed while submerged in a fluid bath that flushes material away, cools the work piece, and acts as a conductor for the current to pass between the electrode and the work piece.

Wire EDM is typically done in a bath of water with a very thin wire serving as the electrode. The wire is slowly fed through the material where the electrical discharges cut out complex contours in the work pieces. This is used in the production of moulds and dies and for improving the surface quality (roughness and metallurgical structure) of machined pieces.

ELECTROCHEMICAL MACHINING

Electrochemical machining (ECM) is a technology that removes metal by anodic dissolution in which a direct current with high density and low voltage is passed from a pre-shaped tool acting as a cathode onto a work piece acting as the anode. The metal on the surface of the work piece surface is dissolved into metallic ions by the de-plating reaction. Dissolved material, gas, and heat are removed from the narrow machining gap by the flow of electrolyte pumped through the gap at a high velocity.

ECM is capable of machining any electrically conductive material with high stock removal rates. Removal rate in ECM is independent of the hardness, toughness, and other properties of the material being machined.

With ECM, complex-shaped parts can be produced by simply moving the tool without rotation in one operation. ECM can remove a defective layer of material and eliminate the flaws inherited by the surface layer from a previous treatment. ECM is the machining method of choice in the case of thin-walled, easily deformable components and for brittle materials likely to develop cracks in the surface layer.

The most common uses for ECM include duplicating, drilling, and sinking operations in the manufacture of dies, press and glass-making moulds, the manufacturer of turbine and compressor blades for gas turbine engines, the generation of passages, cavities, holes and slots in parts, and the like, as well as electrochemical shaping of rotating work pieces and ECM using rotating tool electrode.

ELECTROFORMING

Electroforming is a process that uses electrode position in a plating bath over a base form or mandrel which is subsequently removed. The process synthesizes a metal layer on a surface that has been rendered electroconductive through the application of a paint that contains metal particles. This process deposits a thick layer which can exist as a self-supporting structure if the original work piece is removed. The object being electroformed can be a permanent part of the end product or can be temporary and removed later.

In the basic electroforming process, an electrolytic bath is used to deposit nickel or other electroplatable metals onto a conductive patterned surface. After the plated material is built up to the desired thickness, the electroformed part is stripped off the base. This process allows high-quality duplication of the master form with high product quality with high repeatability and excellent process control.

Electroforming is very effective when requirements call for extreme tolerances, complexity, or light weight. The resolution inherent in the photographically produced conductive patterned base allows finer geometries

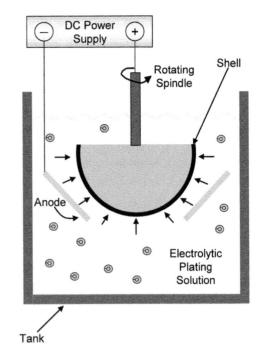

Figure 8-3. Illustration of Electroforming Process

to be produced to tighter tolerances while maintaining superior edge definition with a near-optical finish. Electroformed metal is extremely pure, with superior properties over wrought metal. Multiple layers of electroformed metal can be molecularly bonded together or to different substrate materials.

A wide variety of shapes and sizes can be made by electroforming. Since the fabrication of a product requires only a single pattern, low production quantities can be made economically. In recent years, electroforming has taken on new importance in the fabrication of micro- and nano-scale metallic devices and in producing precision injection molds.

ELECTROCHEMICAL FINISHING

Electrochemical finishing is a process that protects a metallic surface with electrolytic substances. An item to be finished is immersed in an electrolyte and is made the positive electrode. When the electric current passes through the electrolyte, metal is dissolved from the anode surface.

Frequently used electrofinishing processes include passivation, bluing, and parkerizing. Passivation is the process of making a material "passive" in relation to another material. Bluing is a passivation process in which steel is partially protected against rust. Bluing is most commonly used by gun manufacturers, gunsmiths, and gun owners. Bluing is also used for providing coloring for fine metalwork. Parkerizing is a method of protecting a steel surface from corrosion and increasing its resistance to wear through the application of an electrochemical conversion coating.

LASER BEAM MACHINING

Laser beam machining (LBM) is a process that uses a laser beam to melt, burn, or vaporize material away from a work piece. LBM is ideal for making accurately placed holes and can be used to perform precision micromachining on all microelectronic substrates such as ceramic, silicon, diamond, and graphite. Microelectronic micromachining includes cutting, scribing and drilling all substrates, trimming and hybrid resistors, patterning displays of glass or plastic, and trace cutting on semiconductor wafers and chips.

The LBM process can make holes in refractory metals and ceramics and in very thin materials. The laser can scribe, drill, mark, and cut thin metals and ceramics, trim resistors, and process plastics, silicon, diamond, and graphite with tolerances to one micron. Laser beams can be precisely controlled and varied in output and by timeframe. As a result, laser machining is an energy-efficiency technology.

Unlike conventional machining, LBM has no physical contact with the work piece. Materials that are very difficult or impossible to machine using conventional methods can be machined using LBM.

ELECTRON BEAM WELDING

Electron beam welding (EBW) is a welding process in which a beam of high-velocity electrons is applied to the materials being joined. Electron beam processing is used for welding metals, machining holes and slots, and to harden the surface of metals. It is also used for heat treating and melting. Electron beam processing is much faster than conventional welding systems. Other benefits include minimal thermal distortions because the power density and energy input can be precisely controlled, substantially reduced set-up and post-cleaning time, lower labor costs, and the ability to achieve complex and precise heating patterns.

Electron beams operate at temperatures of around 45,000°F. As such, the heat penetrates deeply, making it possible to weld much thicker work pieces than is possible with most other welding processes. The electron beam is tightly focused so the total heat input is lower than that of any arc welding process. Another advantage is higher welding speeds. Metals most commonly welded are stainless steels, super alloys, and reactive and refractory metals. The process can also be used to perform welds of a variety of dissimilar metals combinations.

Electron beam welding is performed using an electron gun enclosed in a large vacuum chamber. A promising emerging technology, called a plasma window, will facilitate non-vacuum electron beam welding in which the plasma window is mounted on the electron gun and maintains the small vacuum area needed to propagate the electron beam. This EB welding technique is more energy-efficiency and can allow for faster production.

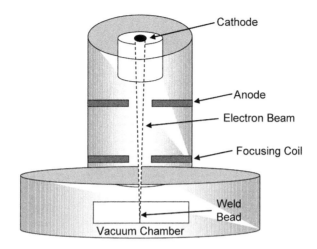

Figure 8-4. Illustration of Electron Beam Welding

LASER BEAM WELDING

Laser beam welding (LBW) joins multiple pieces of metal through the use of a laser. The beam is a concentrated heat source, allowing for narrow, deep welds and high welding rates. Laser processing tends to be faster and has less product distortion compared to conventional techniques.

Laser beam welding has high power density, resulting in small heat-affected zones and high heating and cooling rates. LBW is capable of welding carbon steels, high-strength low-alloy steels, stainless steel, aluminum, and titanium. The weld quality is high.

A type of LBW, laser-hybrid welding, combines the laser of LBW with an arc welding method. This combination allows for greater positioning flexibility since the arc melting method supplies molten metal to fill the joint.

PLASMA WELDING

Plasma welding provides an advanced level of control and accuracy to produce high-quality welds in miniature or precision applications and to provide long electrode life for high-production require-

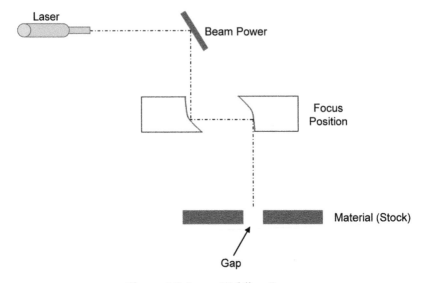

Figure 8-5. Laser Welding Process

ments. Plasma welding is suited to manual and automatic applications and is used in a variety of applications ranging from high-volume welding of strip metal, to precision welding of surgical instruments, to automatic repair of jet engine blades, to the manual welding of kitchen equipment.

Similar to the tungsten inert gas (TIG) process, the plasma arc welding process uses plasma to transfer an electric arc through a tungsten torch. By positioning the electrode within the body of the torch, the plasma arc can be separated from the shielding gas envelope. Plasma is then forced through a nozzle causing an arc. The metal to be welded is melted by the intense heat of the arc. By forcing the plasma arc through a constricted orifice, the welding torch delivers a high concentration of heat.

Several operating modes can be produced by varying bore diameter and plasma gas flow rates.

RESISTANCE WELDING

Resistance welding refers to a group of welding processes (resistance spot welding, resistance seam welding, projection welding, flash welding, upset welding, and high-frequency resistance welding) that apply electricity and mechanical pressure to create a weld. Weld electrodes conduct the electric current to the pieces of metal. Heat or welding temperatures are influenced by the proportions of the work pieces, the electrode materials, electrode geometry, electrode pressing force, weld current, and weld time. In general, resistance welding methods are efficient and cause little pollution, but their applications are limited to relatively thin materials.

Spot Welding

Spot welding is a popular resistance welding method used to overlap metal sheets and simultaneously clamp the metal sheets together and pass current through them. When the current passes through the electrodes to the sheets, heat is generated at the contact points, which melts the metal at that spot. As the heat dissipates throughout the work piece, it cools the spot weld causing the metal to solidify.

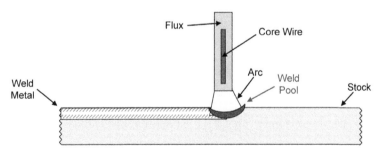

Figure 8-6. Arc Welding Process

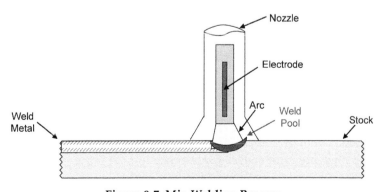

Figure 8-7. Mig Welding Process

The advantage of spot welding include efficient energy use, limited work piece deformation, high production rates, and easy automation. When high shear strength is needed, spot welding is used in preference to more costly mechanical fastening. The overall strength is often significantly lower than with other welding methods.

Seam Welding

Seam welding relies on two electrodes to apply pressure and current to join metal sheets. Wheel-shaped electrodes roll along and feed the work piece, enabling long continuous welds.

Projection Welding

Projection welding is a resistance welding process that produces coalescence of metals with the heat obtained from resistance to electrical current through work parts held together under pressure. The welds are localized at predetermined points by projections or intersections. In projection welding, electrode life is increased because larger contact surfaces are used.

Flash Welding

Flash welding is a resistance welding process that produces coalescence simultaneously over the entire area of two abutting surfaces by the heat obtained from resistance to electric current between the two surfaces. The heat is generated by the flashing and is localized.

Upset Welding

Upset welding is a resistance welding process that produces coalescence simultaneously over the entire area of abutting surfaces or progressively along a joint by the heat obtained from resistance to electric current. Pressure is applied before heating is started and is maintained throughout. The parts are clamped in the welding machine and force is applied bringing them tightly together. High-amperage current is then passed through the joint. When they have been heated to a suitable forging temperature, an upsetting force is applied and the current is stopped. The high temperature of the work at the abutting surfaces with the high pressure causes coalescence to take place. After cooling, the force is released and the weld is completed.

Percussion Welding

Percussion welding is a resistance welding process that produces coalescence of abutting metal pieces using heat from an arc produced by a rapid discharge of electrical energy.

High-Frequency Resistance Welding

High-frequency resistance welding is a resistance welding process that produces coalescence of metals with the heat generated from the resistance of the work pieces to a high-frequency alternating current and the rapid application of an upsetting force after heating is substantially completed. The path of the current in the work piece is controlled by the proximity effect.

UV Curing

Ultraviolet (UV) processing can be used to cure various types of industrial coatings and adhesives, as well as for curing operations in printing and electronic parts applications. UV processing is used extensively in the wastewater industry to treat water and in air treatment systems to purify indoor air. The main applications for UV curing are

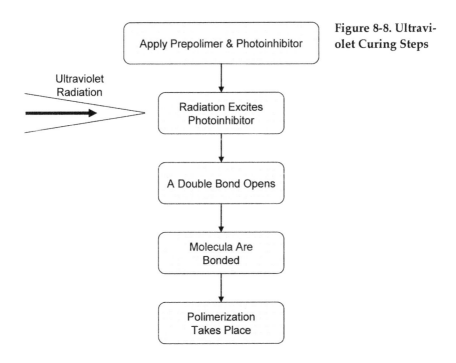

Figure 8-8. Ultraviolet Curing Steps

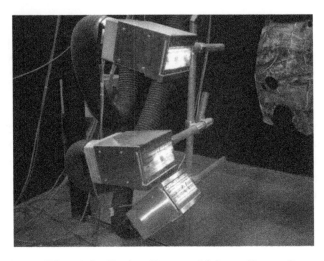

Figure 8-9. Ultraviolet Curing (Source: Alabama Power Company)

coatings, printing, adhesives, and electronic parts.

UV radiation is the part of the electromagnetic spectrum with a wavelength from 4 to 400 nanometers. Applying UV radiation to certain liquid polymeric substances cures the substance into a solid. Curing bonds and fuses a coating to a substrate—sometimes developing specific properties in the coating.

UV radiation is created using UV lamps which typically use mercury lamps or xenon gas arc lamps. UV radiation uses a high-voltage discharge ionized in a tube filled with mercury gas. The discharge can be created by an arc between two electrodes or by microwaves.

UV-based systems have more rapid curing speeds and do not produce as much VOC-based emissions as convection and radiant systems. UV-based systems require special UV-curable coatings and custom-made UV lamp systems for a particular application. UV curing uses around 25% of the energy required by a thermally based system and has instantaneous curing time.

Applications
Wood Coatings

Flat wood stock-UV coatings are widely used for flat wood coatings in Europe. However, in the U.S., the speed of production for those applications is viewed as too slow.

Containers, Metal Decorating

UV radiation curing can be applied to containers, closures, and metal decorating as well as finishes used to coat metal cans and collapsible tubes. It can also be applied to other specialty metal decorative finishes applied by lithography and silk-screen printing on various metal substrates, such as decorative nameplates.

Three-piece cans are coated and printed before construction, that is, on flat stock-idel for radiation curing. Two-piece cans have their interior surface coated with a heat-cured epoxy. The exterior surface is base coated, printed and top coated after assembly. Uniform radiation curing of these exterior steps can be accomplished by rotating the can surfaces. Thus, targets for radiation curing are both the two- and three-piece metal cans.

Motor Vehicles

Automobile manufacturers use conventional waterborne and solvent-based coatings. Other coatings include waterborne primers and powder coatings. In today's auto production, heat-sensitive plastic substrates are spray coated. This necessitates the use of some organic solvent to reduce the viscosity of the coating sufficiently to allow spraying.

All Printing Inks

UV-cured inks and varnishes are used in packaging consumption and printing trade consumption. Printing trade consumption includes folding paperboard cartons used to package food, beverages, soaps and detergents, tobacco, cosmetics, and medicines. Lithographic inks are dried thermally on metal containers and web off-set paperboard carbons.

Infrared Curing

IR coatings are based on thermal evaporation of organic solvents from conventional solvent-based coatings. Thermal energy reaches the wet substrate directly from the IR radiation and from metal reflectors. Pressure-Sensitive Adhesives—This is the single largest application for radiation-cured polymers. Photochemically cured adhesives could potentially replace polymers in the manufacture of pressure-sensitive tapes and labels.

References

"Electrotechnology—Industrial and Environmental Applications," Nicholas P. Cheremisinoff, Noyes Publications, 1996.

"Technology Guidebook for Electric Infrared Process Heating: CMF Report No. 93-2," EPRI, Palo Alto, CA: 1993. CR-102785.

"Overview of the Cement Industry," Portland Cement Association (www.cement.org/basics/cementindustry.asp), 2009.

"Electron-Beam Heating of Metals," G. Goretik and S. Rozin, *Journal of Engineering Physics and Thermophysics*, 1972.

"Electroforming," www.dalmar.net/electroforming.htm, 2010.

"Manufacturing Processes—Laser-Beam Machining (LBM)," Engineers Handbook.com, 2006.

"Welding Principles and Applications," L. Jeffus, Delmar Cengage Learning, 2008.

"Electric Process Heating—Technologies/Equipment/Applications," M. Orfeuil, Battle Press, 1987.

Chapter 9

Beneficial Commercial Building Uses of Electricity

As in the other sectors, there are ample opportunities for the increased penetration of beneficial new uses of electric devices and appliances in the commercial building sector. In almost every case, these devices are integral building energy technologies. In fact, the most important areas of fossil energy use which can be converted to electricity with lower overall energy use and lower CO_2 emissions in commercial buildings are space conditioning, water heating, and food service. This chapter explores those opportunities.

The following are descriptions of the most important beneficial new uses which can be applied to commercial buildings.

SPACE CONDITIONING

Space-conditioning systems afford the consumer healthy living and working environments that enable productivity and a sense of well being. Space conditioning is the largest end user of electricity in commercial buildings. The primary purposes of space conditioning are to heat, cool, dehumidify, humidify, purify air and provide ventilation. Space conditioning equipment includes humidifiers, dehumidifiers, resistance heaters, heat pumps, electric boilers, and various controls. Innovations in technologies related to space conditioning may have a substantial effect on how electricity is used in the future.

Heat Pumps for Space Conditioning

The electric heat pump is the primary device that can shift commercial markets from natural gas to electrification at a rapid pace, coincidentally reducing the nation's total resource needs and substantially reducing overall emissions. Efficient electric power generation, coupled with an electric heat pump, uses less resource energy and emits less CO$_2$ than high-efficiency gas heating coupled with electric cooling. (EPRI CU-7441)

Using electric heat pumps in the commercial market leads to substantial advantages. Commercial buildings are often comprised of multiple zones, some requiring heating while others requiring cooling. In addition, even during a cooling phase, domestic hot water heating requirements can be met by recovering heat from the cooling (refrigeration) circuit.

For example, water-loop heat pump systems use an un-insulated water loop as the source or sink for multiple heat pumps. In medium- to large-sized commercial buildings with significant internal thermal loads, heat recovery from the individual units, using the water loop, reduces boiler and cooling tower operation, thereby reducing building energy costs.

Heat recovery heat pumps are heat recovery chillers with a heat pump refrigeration circuit that boosts waste heat to a useful temperature. These units are used for space heating and domestic water heating, especially when water temperatures of 120°F to 130°F are required. The

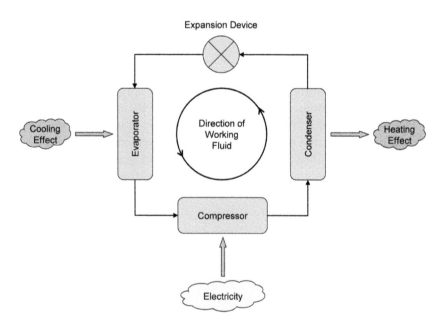

Figure 9-1. Heat Pump Operation

units can be retrofitted to existing systems.

Another commercial system that has a sizeable technical potential is the heat recovery electric chiller. This system supplies heat for space heating or domestic water heating by the heat normally rejected to a cooling tower. Such a unit also supplies chilled water for air conditioning. The best economics for this system are achieved when the water heating temperature requirement is 110°F or less, or when the chiller is used as a pre-heater for another water heating unit.

Because commercial technologies overlap, it is necessary to segment the market further to avoid double accounting. The following four secondary segments are used:

- Buildings with chillers and hydronic heating systems are candidates for heat recovery chillers and heat recovery heat pumps. Water-loop heat pumps and air-source heat pumps are not considered for these buildings to avoid redundant cooling capacity.

- Buildings with chillers but without hydronic space heating systems are candidates for heat recovery with respect to water heating only. The presence of chillers precludes the use of heat pumps for space heating or cooling.

- Buildings with other types of cooling equipment (non-chiller systems) and with hydronic heating systems are candidates for water-loop heat pumps.

- Buildings with other types of cooling equipment (non-chiller systems) and with non-hydronic heating systems are candidates for air-source heat pumps.

Water-Loop Electric Heat Pumps

Water-loop electric heat pump systems have the following four components:

- Distributed heat pump units in the space (perimeter and interior zones).

- Cooling tower for heat rejection when there is insufficient heating demand.

- Boiler for adding heat to the loop when heating load is greater than heat pump output.

- Circulating pumps.

The advantages of these systems are reasonable first cost, design flexibility, smaller space requirement for mechanical rooms, and potential tenant metering.

Retrofit is plausible where there is an existing hydronic system. Otherwise, major piping is required. Retrofit is unlikely where there is an operable chiller. Hence, the commercial market for water-loop heat pumps is rather narrow, consisting of buildings with hydronic heating systems (boilers) but with no cooling or with ducted or distributed cooling systems. This is less than 20% of the existing stock.

In new construction, water-loop heat pump systems are increasingly popular, based largely on factors other than energy efficiency. Albeit, they are extremely efficient. Application is limited to larger buildings (over 25,000 square feet), accounting for about 40% of new construction. On the cooling side, the competition is chiller-based systems for the larger buildings, and packaged units for low-rise and medium-sized buildings. Chillers still maintain a significant market share in larger buildings and a significant share in the medium-sized category.

Water-loop heat pump efficiency depends on the operating mode:

- When there is simultaneous cooling, the distributed heat pumps have a coefficient of performance (COP) of more than 4. This means that 0.25 Btu of electricity is required for each Btu of heat added to the space.

- When heat from a resistance boiler is required, a resistance system results, with an overall COP of 1.0. For every Btu of heat added to the space, the distributed heat pump supplies 0.25 Btu, and the remote boiler supplies 0.75 Btu.

- When heat from a fossil-fueled boiler is required, a dual-fuel system results. For every Btu of heat added to the space, 0.25 Btu of electricity is required and 1.0 Btu or more of fossil fuel is required (assuming a boiler efficiency of 0.75). In this case, the heat pump supplies 20% and the boiler 80% of the heating requirement.

Additional assumptions used in these conclusions include:

- For the computations, water-loop systems are assumed to be in simultaneous mode about 67% of the time and in backup mode about 33% of the time.

*   Averaged over an annual cycle, electric boiler input is about 0.5 Btu per delivered Btu, with an additional 0.25 Btu of electricity required by the distributed units. This yields a total requirement of 0.75 Btu/Btu and a seasonal COP of 1.33.

*   With a fossil boiler, average input requirements per Btu of heat delivered are about 0.25 Btu of electricity and 0.67 Btu of fuel.

In this discussion, all water-loop systems are assumed to have a boiler as the primary heat source. Although it is technically feasible, heat pumps are not considered for the central plant for these systems. This approach would be excessively costly and has rarely, if ever, been used. Also, heat recovery from the water loop is a possibility and is a major advantage of the heat pump.

Commercial Building Air-Source Electric Heat Pumps

Air-source electric heat pumps can be applied in a wide variety of commercial buildings. This category covers the following:

*   Packaged terminal heat pumps—low-capacity units mounted in through-the-wall sleeves.

*   Unitary heat pumps including single-package units, which are usually roof-mounted, and split systems.

*   Dual-fuel heat pumps including an electric heat pump with supplemental heating provided by an internal gas furnace rather than backup resistance elements.

Air-source heat pump retrofit is plausible for most small- and medium-sized buildings that do not have hydronic heating systems. (Buildings with hydronic systems are covered in the water-loop heat pump category.) In the commercial market, primary competition to these heat pumps includes conventional packaged equipment and packaged terminal equipment. Add-on units are also feasible, and these are assumed to function like dual-fuel equipment.

Air-source heat pumps provide a feasible and increasingly popular option in new commercial construction in the small and medium low-rise categories.

VARIABLE CAPACITY AIR CONDITIONERS AND
HEAT PUMPS FOR COMMERCIAL BUILDINGS

Multi-split heat pumps have evolved from a technology suitable for residential and light commercial buildings to variable capacity or variable refrigerant flow (VRF) systems that can provide efficient space conditioning for large commercial buildings. VRF systems are enhanced versions of ductless multi-split systems, permitting more indoor units to be connected to each outdoor unit and providing additional features such as simultaneous heating and cooling and heat recovery. VRF systems are very popular in Asia and Europe and, with an increasing support available from major U.S. and Asian manufacturers are worth considering for multi-zone commercial building applications in the U.S.

VRF technology uses smart integrated controls, variable-speed drives, refrigerant piping, and heat recovery to provide products with attributes that include high energy efficiency, flexible operation, ease of installation, low noise, zone control, and comfort using all-electric technology.

Manufacturer and case study information from outside the U.S. show systems that are extremely cost-effective, but results, again, depend on specific application features. Energy savings are achieved, ranging from 10% to 60%, depending on climate and the type of system displaced, among other factors. Initial costs are also typically higher with payback periods dependent on energy savings. Examples of application and operation costs outside the U.S. include the following (EPRI 1016288):

- Anecdotal information exists showing savings of 38% in a side-by-side comparison with a rooftop variable air volume (VAV) installation, but this study compared a new VRF system to the existing rooftop VAV system. Simulation results for a Brazilian climate showed savings of over 30% in summer and over 60% in winter. Based on this study, savings of 5% to 15% were suggested to be more likely in the United States.

- A modeling study conducted with a version of EnergyPlus that was modified to simulate VRF systems showed that, in a ten-story office building in Shanghai, a VRF system saved more than 20% of the energy compared to a VAV system and more than 10% compared to a fan-coil-plus-fresh-air system.

- Information obtained in discussions with one manufacturer suggests that the company's VRF systems could save up to 30% to 40% of the energy used by a chiller-based system for a 200 ton cooling system for a generic commercial building. The same data set indicates that the installed cost of a VRF system will be about 8% more than a water-cooled chiller and 16% more than an air-cooled chiller. Combining these energy use and installed cost projections provides an estimated payback period of about 1.5 years for the VRF compared to an air-cooled chiller and two-thirds of a year compared to a water-cooled chiller.

- Data from another VRF manufacturer compared installation and operating costs for a set of 14 branch buildings in Central/Northern Italy, where a chiller/boiler system was installed in 7 of the buildings and a VRF system was installed in the other 7 buildings (designed to handle heating down to -20°C [-4°F]) in 1998. The VRF systems used 35% less energy and had 40% lower maintenance costs for the period studied. As suggested by other manufacturers, the equipment costs for the VRF systems were higher than the equipment costs for the chiller-based systems, but this was offset by lower installation costs for the VRF systems.

Initial applications of VRF in commercial buildings have included building add-ons such as a new data center and situations where spot cooling is needed. Historical buildings have benefited from the minimum alterations needed for the addition of a VRF system. Retrofit situations where air conditioning may be an addition/upgrade to the space can be good applications of ductless systems since additional duct work and conditioning needed for ventilation can be minimized with VRF systems compared to ducted systems.

Other applications well suited to VRF systems include cases in which there is an advantage to delivering personalized, compartmentalized comfort conditions, such as:

- Office buildings with multiple individual offices
- Strip malls
- Hotels and motels
- Hospitals and nursing homes
- Banks
- Schools

Most offices and strip malls have occupants with different space conditioning requirements, making the zoning opportunities of VRF attractive for these applications. Hospitals and nursing homes can be good candidates since the VRF system makes it easy to avoid zone-to-zone air mixing. And banks have favored the system for security because the egress paths into the bank are minimized due to either the elimination of duct work entirely or the use of minimal smaller-diameter ductwork. Even in schools, which because of high occupancy often have a 100% outside air requirement, VRF units can be used (often with heat recovery ventilators) to meet heating and cooling requirements.

Commercial Electric Heat Recovery Chillers and Electric Heat Recovery Heat Pumps

Electric heat recovery chillers and electric heat recovery heat pumps provide domestic hot water and space heating when the cooling load is being met.

- Heat recovery chillers supply useful heat by recovering all or part of the heat normally rejected to a cooling tower. Two methods are used to recover heat:
  — A second water-cooled condenser is added to the system that is operated in parallel with the normal cooling tower condenser, or
  — A heat exchanger is placed between the condenser and the cooling tower.
- Heat recovery heat pumps use a heat pump instead of a condenser or heat exchanger, adding the ability to boost recovery temperatures.

Heat recovery chillers using a second condenser can be difficult to retrofit into an existing system. Heat recovery heat pumps, however, are easily retrofitted to existing systems.

Heat recovery chillers and heat pumps are viable in all buildings with chillers and hydronic heating. Selection of the specific heat recovery technology depends on the following:

- Heat recovery chillers are best suited to lower-temperature applications (100°F or less), or they can be used as a pre-heater for conventional water heating equipment.

• Heat recovery heat pumps are well suited to applications where heating is required at higher temperatures.

Electric Infrared Space Heating
Heat is transferred in three ways: convection, conduction, and radiation. In most space heating systems, convection and conduction are the principle heat transfer mechanisms. Although these heaters are referred to as "space heaters," they do not directly heat the space; they heat the objects in the space which, in turn, eventually heat the space. The term "infrared space heating" is used to distinguish comfort heating applications from IR process heating.

Electric IR heaters have two basic components: an IR heating element and a reflector. The IR heating element is composed of a resistor material (or radiator) that gives off electromagnetic energy in the IR portion of the spectrum when excited by an electric current. The resistor material is partially enclosed in the reflector, a fixture that reflects the radiation toward the people to be heated. Resistor materials which can be used include tungsten wire in a quartz tube, nickel chromium alloy in a quartz tube, tungsten wire in a reflector lamp, and nickel chromium alloy in a metal rod. In space heating applications, the IR radiation is normally directed toward the people in the area. However, the radiation also strikes objects, such as the floors, walls, equipment, and furnishings. These objects then retransmit the heat they receive, through secondary convection, conduction, and radiation. In this way, IR heaters can be used to warm the air in a room to a set temperature, much like a conventional heating system.

Tungsten wires in quartz lamps and reflector lamps operate at filament temperatures of about 4050°F and radiate energy in the "near-infrared" portion of the spectrum. These lamps have the added advantage of providing visible light of approximately 8 lumens per watt. This light can help illuminate work areas. The potential downside is that when heating is not needed, the extra light is not provided. Other lamp elements, such as metal sheath, open wire, and ribbon elements, operate between 1200°F and 1800°F and emit in the "far-infrared" portion of the spectrum.

IR heaters are used in a variety of applications including golf driving ranges, storage rooms, fire station garages, loading docks, covered walkways, warehouses, and shops, outdoor restaurants, hotels and motels, shopping centers, and store entrances. The most cost-effective applications are in areas exposed to the outdoors and/or areas that require high ventilations rates.

An IR system can be designed to maintain an air temperature of over 70°F in an enclosed room. In this case, the overall efficiency of the system approaches 100%, but as little as 50% of the radiated energy actually reaches the people or objects. The ideal application is an area maintained at 50°F in which IR heaters are used only to warm people. An example is a large warehouse watched over by a stock person near the main door. IR heaters could be installed where the stock person spends the most time, providing the stock person heating comfort irrespective of the overall warehouse temperatures.

IR lamps and fixtures are available in a variety of shapes and sizes. They are normally hung from or attached directly to the ceiling, in a manner similar to a lighting system, with careful attention to the maximum height of forklifts, trucks, cranes, etc., that operate in the area. The IR system designer determines the desired energy levels for the different parts of the facility and then estimates the equipment wattage required to produce the desired energy levels. The fixtures are typically available in 120-, 240-, 277-, and 480-volt systems. Simple switches, timers, and thermostats are used to control fixture output.

The lamp efficiency depends on the material of the resistor or radiator. Clear quartz lamps have an efficiency of about 96%. Tungsten wire and quartz tubes have efficiencies of 60 to 80%. Metal rods have lamp efficiencies as low as 50%. The overall efficiency of the IR system depends on the type of IR element in the lamp, the absorptivity of the people and the objects near the lamp, and the efficiency of the fixture (including the reflectivity of the reflector and the directional efficiency of the fixture). Other factors to consider in selecting an IR element include amount of visible light output, time required to develop full output, vibration resistance, and color of light. The life expectancy of any IR lamp is about 5000 or more hours.

Ozonation of Cooling Tower Water

Air conditioners, heat exchangers, power generators, and other large machines generate heat while operating wet. Cooling tower systems are often used to absorb the heat discharged from these units by circulating water through and around them. The cooling tower itself removes the heat absorbed by the water so the water can be reused.

Without treatment, cooling tower water will develop and breed

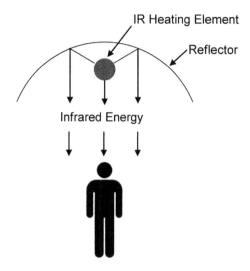

Figure 9-2. Electric Infrared Space Heating Illustration

microorganisms. A combination of chemicals are typically used to treat cooling tower water; research has identified ozonation as a potential alternative. Ozone is a powerful antioxidant; it readily attacks and breaks down exposed carbon bonds, leaving no residue. Ozone itself breaks down into oxygen, so there is no chemical waste product. Ozonation systems are a proven success in water and wastewater treatment.

Most wet cooling towers use an evaporation process to remove the heat absorbed by the cooling water. In this process, heated water flows into the cooling tower and is then sprayed as small droplets onto wet decking surfaces. After cooling, the droplets fall into a cold water basin and are pumped back into circulation. Some of the droplets, of course, evaporate and exit as steam; therefore, the water supply must be replenished periodically. It also must be treated regularly to prevent algae growth and scale buildup.

Conventional cooling tower systems use multiple chemicals for water treatment, but this is problematic. If not enough chemicals are used, scale buildup causes inefficiency in energy use and frequent system shutdown for cleaning. Excessive use of chemicals is costly and can lead to metal corrosion and unusually high chemical concentrations in the wastewater and atmosphere. Over time, the continual addition of chemicals plus water loss through evaporation results in water that is saturated with chemicals and debris, making emptying and refilling with a new water supply necessary.

Using ozone solves many of these problems. Ozone acts as a biocide by oxidizing the cellular structure of microorganisms in water. It inhibits scale deposits and limits spontaneous oxidation in metals by forming ferroso-ferric oxide (Fe$_3$O$_4$) with the metal. In addition, no residual chemicals are left in the water.

Because it is highly reactive, ozone must be generated on-site as needed and, for best economics, injected in concentrations proportional to the organic and biological demand of the water being treated. Ozone can be generated by corona discharge or by ultraviolet (UV) excitation. In the corona discharge method, air or oxygen is passed between two electrodes, and a corona discharge is generated by applying high voltage to the electrodes. In the UV excitation process, ozone is generated photochemically by passing air through a path irradiated with UV light. A portion of the air dissociates and recombines to form ozone. The corona discharge method is the most commonly used. The UV excitation method is limited in application because of its high-energy requirements (20 kWh per pound of ozone produced) and relatively low ozone production rate.

Ultraviolet Disinfection of Air

Filters and proper ventilation are an effective means of removing dust, pollen, and other airborne particles from indoor air but are not effective against microorganisms such as bacteria and viruses. These germs are most prevalent where people congregate—in hospitals, offices, hotels, and stores. Since the indoor air is often re-circulated in these places, the

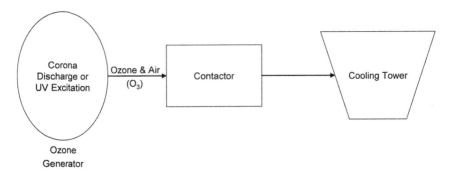

Figure 9-3. Ozonation of Cooling Tower Water

chances of transmission of an infection through coughing, sneezing, and/ or talking are great. To prevent transmission of airborne diseases such as tuberculosis (TB), many institutions (office buildings, shopping centers, hospitals, and hotel/motels) are installing air sanitization units.

Certain ultraviolet (UV) wavelengths are capable of sterilizing airborne germs. The UV spectrum lies between X rays and the visible spectrum (180 to 400 nm). Within this range, UV rays are further divided into three spectrums: the long-wave UV-A (315 to 400 nm) rays are generally used in tanning salons; the medium-wave UV-B (280 to 315 nm) rays are found in sunlight and cause skin cancer and tanning; and the short-wave UV-C (<280 nm) rays are used for disinfection of airborne germs. UV-C rays cannot penetrate the skin to cause cancer, nor can they reach the lens of the eye to produce cataracts. UV-C can, however, penetrate the cell wall of a microorganisms and cause photo-chemical breakdown of its DNA.

The peak UV absorption efficiency for DNA (thus, the optimal range for destruction of airborne germs) is between 250 to 260 nm. UV lamps are an ideal source because the energized cathodes are designed to emit rays at 253.7 nm. These lamps are similar to fluorescent tube lamps except the tubes are made of quartz glass and the tube's inner surfaces are not coated with phosphor. There are two ways to install UV lamps for air disinfection: as ceiling- or wall-mounted fixtures, or within ventilation ducts.

The first method, known as overhead or upper-air irradiation, mounts the light fixture to either the ceiling or a wall, depending on the height of the room. Overhead installations are best for ceilings that are at least nine feet high, so the fixtures extend no lower than seven feet. This is necessary so people do not bump their heads on the fixtures or look directly into the UV rays. These fixtures should be shielded on the bottom and partially on the side so the UV rays radiate upward or out the side, but not down, to protect people from

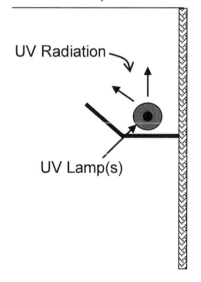

Figure 9-4. Typical Wall Installation of a UV Unit

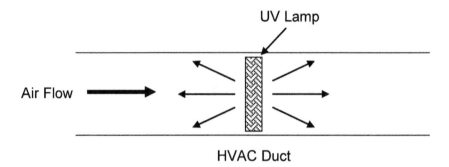

Figure 9-5. Illustration of a UV Installation in an HVAC Duct

direct exposure.

The second method is to install UV lamps inside the ventilation ducts of a re-circulating system. Very high levels of UV light can be used in this system because the ducts shield people from direct exposure to the rays. Since the effectiveness of UV rays increases with duration of exposure, it is best to install the lamps where airflow is slowest, perhaps behind the filters. It is also ideal to install the lamps perpendicular to the airflow so that the light radiates in both directions throughout the length of the air duct. If the lamp is installed at the bottom of the duct, light can only radiate upward at the air that passes over the lamp. The system should also be designed with an inspection window to allow periodic checking of the lamps. Note that the lamp must automatically shut off when the window is opened to protect maintenance personnel from direct contact with the UV beams.

ADVANCED HEATING AND COOLING TECHNOLOGIES

Thermotunneling-based Cooling

Thermotunneling-based cooling systems would potentially replaced the standard vapor compression cycle in cooling and refrigeration systems. The system has the potential to increase energy efficiency by 1.5 to 2 and has no moving parts or gas refrigerants. In thermotunneling, electrons carry heat from one side of an insulated gap (the cathode) to the other side of the gap (the anode). The cathode side is cooled, and the anode side is the heat sink. Electrons are transferred by tunneling, whereby electrons jump across the gap. The gap must be very small (1 to

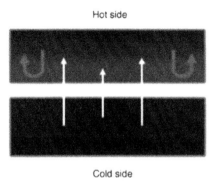

By introducing a gap, the return path for
conducted heat (red arrows) is eliminated,
making Cool Chips extremely efficient.

Figure 9-6. Thermotunneling (Courtesy of Cool Chips LLC)

10 nanometers), and a voltage bias must be applied to encourage flow in
the desired direction. The technology is applicable to cooling processes in
a variety of applications including space cooling, refrigeration, and cooling. It is an alternative to the vapor compression cycle across all sectors.

As shown in Figure 9-7, the electrons that move across the physical
gap in thermotunneling devices cannot easily return. As a result, conducted heat is eliminated. There are certain materials that emit electrons
more easily, rendering them in thermotunneling devices.

Active Magnetic Regenerative Cooling

Active magnetic regenerative (AMR) cooling technology relies on
the magneto-caloric effect exhibited by materials when exposed to a
magnetic field. The electron spins in these materials align under the influence of a magnetic field. This ordering of the electron spins results in
a release of energy in the form of heat, and the material heats up. When
removed from the magnetic field, the spins return to their random order,
absorbing heat in the process, and the material cools down.

The magneto-caloric effect is being investigated as a method to
cool without the use of chemical refrigerants as an alternative to the
vapor compression cycle. Figure 9-8 illustrates the principles of AMR
cooling. In this process, a magneto-caloric material is cycled through
magnetized and demagnetized states, and the heat is vented off to yield

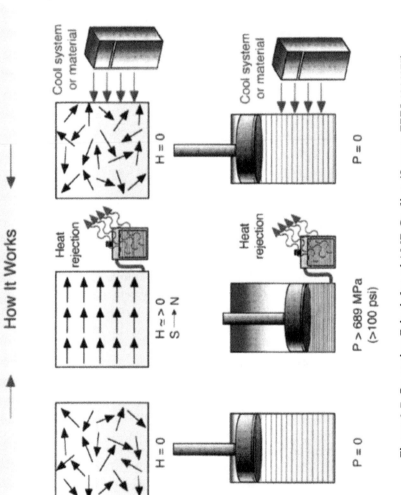

Figure 9-7. Operating Principles of AMR Cooling (Source: EPRI 1016875)

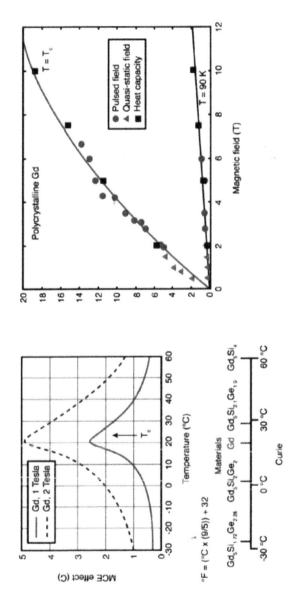

Figure 9-8. Near-Ambient Magnetic Cooling (Source: EPRI 1016875)

a net cooling effect. The effectiveness of the technology is a function of the properties of the material and the source of the magnetic field. Researchers have developed a prototype magnetic refrigerator based on the material gadolinium and a permanent magnet.

New giant magnetocaloric effect (GMCE) materials with much larger MCEs at lower applied magnetic field changes has resulted in the potential for increased efficiency. Gadolinium (Gd) and its alloys are currently the best available materials for magnetic refrigeration near room temperature, as illustrated in Figure 9-9. The MCE is most intense at the Curie temperature (Tc), and higher magnetic fields intensify the MCE. Since magnetic forces increase with increasing magnetic field, it is critical to balance forces. It is necessary to get the heat transfer fluid flow to match the magnetic forces.

AMR would provide an environmentally friendly alternative to hazardous chemical refrigerants used in the conventional vapor compression cycle. In addition, the technology has the potential to be more energy efficient than conventional cooling systems, and the technology requires few moving parts.

The technology is potentially applicable to a variety of cooling and refrigeration applications in the residential, commercial, industrial, and transportation sectors including refrigeration, HVAC systems, and household appliances.

Thermoelectric Cooling

Thermoelectric materials are capable of generating electricity from thermal gradients, or from generating thermal gradients from electricity. Figure 9-10 illustrates thermoelectric modules used for refrigeration and power generation. The capability of thermoelectric materials to generate thermal gradients from electricity could allow for compressorless refrigeration and air conditions. Thermoelectric cooling and heating sometimes incorporates the *Peltier effect*.

Thermoelectric materials are characterized by a dimensionless figure of merit (FOM), ZT:

$$ZT = \alpha^2 T / \rho \kappa$$

Where $\alpha = \Delta V / \Delta T$ (the Seebeck coefficient), V is voltage, T is absolute temperature, ρ is electrical resistivity, and κ is thermal conductivity.

Conversion efficiency for cooling can be conveniently expressed in

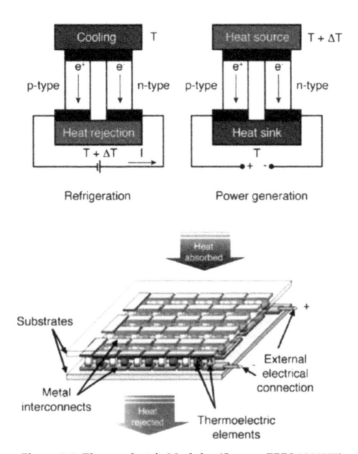

Figure 9-9. Thermoelectric Modules (Source: EPRI 1016875)

terms of only ZT and temperature T. The best thermoelectric materials currently have less than 5% conversion efficiency, which corresponds to a ZT of ~1. Because of the lower conversion efficiency, thermoelectric materials are presently used only in niche applications, such as automobile seats, small portable refrigerators, and wine coolers. Recent advancements in nano-based thermoelectric materials with ZT values exceeding 2 have resulted in improvements in energy efficiency.

Series Desiccant Wheel for Improved Dehumidification
      This technology, which is also referred to as the Cromer Cycle, uses a "series" desiccant wheel to improve the dehumidification ability of a cold coil. The desiccant wheel is installed in the air handling unit

(AHU) where it boosts the capacity of the cooling coil, increasing moisture removal. Moisture from the air exiting the cooling coil is captured in the top part of the wheel and then transferred to return air that goes through the lower part of the wheel. The wheel transfers the latent load from air exiting the cooling coil and passes it through the cooling coil again. Very little sensible load is transferred in the process. The wheel rotates at about eight revolutions per hour.

The series desiccant wheel can enable the air leaving the AHU to have a lower dew point than the air leaving the coil, thus improving the dehumidification ability of the coil. The technology can increase energy efficiency by reducing the energy required for reheat and cooling.

The technology is applicable to commercial and industrial buildings requiring humidity control such as hospital operating rooms, schools, dry storage, museums and courthouses.

**Condenser Heat-Reactivated Desiccant for
Improved Dehumidification**

The condenser heat-reactivated desiccant technology dehumidifies ventilation air with a desiccant that is regenerated with recycled condenser heat. This technology is a hybrid technology that merges the benefits of desiccant dehumidification with direct expansion (DX) air conditioning. It enables humidity and temperature to be controlled independently. It differs from conventional DX systems in that under moderate humidity conditions, conventional systems must overcool the air in order to dehumidify it, and the air is then reheated to the appropriate temperature. This requires considerable energy. The HCU employs a desiccant rotor to dehumidify the air. The desiccant is then regenerated with recycled heat from the condenser, rather than with a separate gas heater.

A major advantage of the HCU is that it controls humidity separately from temperature. Conventional air conditioning relies on the cooling process to also dehumidify the air. When the latent load (humidity) is high relative to the sensible load (temperature), conventional systems may not adequately dehumidify the air, or they may cause the temperature to be low, both of which cause occupant discomfort and complaints. The HCU unit is more energy-efficient than traditional DX-only systems. Better humidity control helps protect building owners and operators from the liabilities associated with mold growth.

Photocatalytic Oxidation of Air Pollutants

Photocatalytic oxidation (PCO) technology has the potential to improve upon some of the key attributes important to meeting air quality. PCO uses ultraviolet (UV) light and a photocatalytic semiconductor material to promote the formation of highly reactive chemical species. The chemical species, in turn, oxidize pollutants such as microorganisms, allergens, and VOCs.

The majority of products available in the indoor air purification market utilize titanium oxide (TiO_2) as the photocatalyst material. The oxidizing species turn VOCs into harmless compounds such as carbon dioxide (CO_2) and water (H_2O). In addition, as microorganisms pass through the reactor, their cells rupture, leading to inactivation of the organisms.

PCO systems are applied to commercial and industrial indoor air purification applications in areas where microorganisms are found in higher levels and/or when microorganisms destruction is essential for the health of building occupants. In particular, they are applicable to health care facilities, schools, theaters, and office buildings.

Air Filtration with Dielectric Barrier Discharge

Hybrid air purification systems coupling air filtration with dielectric barrier discharge have the potential to improve upon some of the key attributes important to meeting indoor air quality needs. This family of technologies is capable of destroying particulates captured in filter media by employing a non-thermal plasma generated by dielectric barrier discharge—in essence, the plasma sterilizes the filter.

Dielectric barrier discharge processes employ plasma, which is often referred to as the fourth state of matter, in the destruction of a wide variety of airborne contaminants including viruses, bacteria, volatile organic compounds (VOCs), heavy metals, and other toxic air pollutants. This is a specific type of advanced oxidation process. NTP technologies generate energized electrons that collide with atoms and molecules in the gas stream to create highly reactive chemical species, more controllable compounds.

There are several types of NTP reactor technologies including electron beam, corona discharge, pulsed corona discharge, dielectric barrier discharge, and flow stabilized discharge.

Electron Beam for Air Purification

Electron beam technology has the potential to improve upon some of the key attributes important to meeting indoor air quality during the next 15- to 20-year timeframe. Electron beams can break down gaseous contaminants found in indoor air into simpler and less harmful chemicals.

The electron beam technology is non-thermal plasma. For non-thermal plasma technologies, discharge electrodes within the gas stream generate electrons. For the electron beam method, electrons are generated externally from the gas. An electron beam system is comprised of four major components: (1) electron beam emitter, (2) high-voltage supply and cable, (3) computer and controller, and (4) shielding.

The electron beam emitter generates high-energy electrons that break down VOCs and odors into simpler and less harmful chemicals. The newer electron beam units use solid-state power supplies and are hermetically sealed (no vacuum pump is required), resulting in compact and modular units that can easily be scaled and integrated into any environment.

Primary applications are air sterilization (airplanes, clean rooms, hospitals, anti-microbial warfare, packaging), destruction of VOCs, destruction of odors, and elimination of nitrogen oxides (NO$_x$) and sulfur oxides (SO$_x$) from exhaust and flue gases.

SPACE CONDITIONING AND/OR WATER HEATING USING CARBON DIOXIDE (CO$_2$) REFRIGERATION CYCLE

The carbon dioxide (CO$_2$) refrigeration cycle, or the Shecco cycle, uses CO$_2$ as the working fluid in a heat pump for heating or cooling. CO$_2$ is advantageous because it has a very low boiling point, and therefore, can absorb heat from the surroundings and evaporate even at very low air temperatures. CO$_2$ is more environmentally friendly than chemical refrigerants such as chlorofluorocarbons (CFCs) and hydrochlorofluorocarbons (HCFCs) used in the conventional vapor compression cycle. The CO$_2$ cycle can replace fossil fuel-fired heaters. The technology can be used for simultaneous space conditioning and water heating.

CO$_2$ is non-flammable and non-toxic. It was already in use in the 1930s in refrigeration and air conditioning equipment, but it was replaced in the 1940s and 1950s by synthetic refrigerants. Because of the phase-outs of high ozone depletion potential and high greenhouse warming potential refrigerants (CFCs and HCFCs), there is a renewed interest in CO_2 as a refrigerant.

The use of CO_2 as a refrigerant was reintroduced in the 1980s by Professor Lorentzen and his colleagues at the Norwegian Technical University Foundation for Scientific and Industrial Research (SINTEF). Their efforts have resulted in the CO_2 technology that is now commercially available in heat pump water heaters, mobile air conditioners, vending machines, coolers, and supermarket refrigeration systems in Japan and Europe. R&D around the world is furthering the use of CO_2 as a refrigerant in additional residential and commercial end-use applications, including integrated space conditioning (or refrigeration) and water heater units.

CO_2 has a low critical temperature of 31.1°C (88.0°F). Its very low boiling point allows it to absorb heat from the surroundings and evaporate even at very low air temperatures. CO_2 vapor compression systems operating at normal ambient temperatures will work close to or even above the critical pressure. CO_2 systems will then operate partly below and partly above the critical pressure. In the CO_2 transcritical cycle, heat is rejected at a supercritical pressure, resulting in a large refrigerant temperature glide.

The CO_2 transcritical cycle is well suited for large temperature cycles, such as those encountered in water heating applications where CO_2 heat pump water heaters can deliver hot water at high temperatures. At near-critical pressure, all or most of the heat transfer from the refrigerant takes place through cooling of the compressed gas. As such, the heat-rejecting heat exchanger is often called a gas cooler instead of a condenser. The refrigerant can be cooled to a few degrees above the entering coolant temperature, resulting in a high coefficient of performance (COP).

CO_2 can be used as a refrigerant in space conditioning and refrigeration. There are three primary ways to use CO_2 as a refrigerant:

- **Direct refrigerant**. When CO_2 is used as a direct refrigerant, the high-pressure side of the system operates at pressures above the critical pressure—so-called *transcritical operation*. Transcritical operation works well for water heating. However, it typically cannot compete in terms of efficiency when used for space conditioning because the vapor compression air conditioning cycle is theoretically more efficient at pressures well below the refrigerant's critical point. The use of economized screw compressors can significantly improve the efficiency of transcritical CO_2 systems at lower temperatures.

- **Indirect volatile secondary refrigerant.** When CO_2 is used as an indirect volatile secondary refrigerant, it is circulated to evaporators. In the evaporators, it extracts heat and returns to a sump, where it condenses by the action of a conventional refrigerant. In this system, no compressor is required for CO_2. Since the mass flow of circulating CO_2 is much less than that of non-volatile secondary refrigerants, such as glycol, the pumping power is greatly reduced. These systems can be oil-free, which, in turn, improves the heat transfer coefficient. The use of CO_2 as a volatile secondary refrigerant has been applied in more than 100 refrigeration systems at supermarkets and cold store facilities in Europe.

- **Low-temperature refrigerant in cascade systems.** In cascade systems, CO_2 is used as the low-temperature refrigerant, and ammonia is typically used as the high-temperature refrigerant. There are two important advantages associated with the use of CO_2 in this way. First, ammonia is confined to the equipment room. Second, it overcomes the problems caused by the very high specific volume of ammonia vapor at low temperatures, which require large piping sizes. Several cascade systems have been installed in commercial applications in Europe, Australia, and the United States.

The following are the primary merits of CO_2 as a refrigerant:

- **Important alternative refrigerant.** CO_2 is an environmentally benign alternative to many conventional refrigerants. It has become the German car manufacturers' choice for replacement of HFC-134c in mobile air conditioners. It is expected that CO_2 will become a viable replacement for HFCs in residential and commercial space-cooling and refrigeration applications.

- **Smaller components.** Because of the higher pressure levels in CO_2 systems, smaller piping is needed.

- **High compressor efficiency.** Due to the smaller compressor displacement and lower compressor ratios, the compressor efficiency is higher.

- **Energy savings**. CO_2 systems have the potential to save energy through improved compressor efficiency and reduced pumping energy use.

- **Reduced capital costs**. Because CO_2 systems use smaller components, the capital costs can often be reduced.

- **Widely applicable**. There are numerous possible end uses of CO_2 as a refrigerant, including heat pump water heaters, heat pump dryers, residential heat pumps, residential air conditioners, mobile air conditioners, vending machines, commercial refrigeration, and commercial space conditioning.

The following are the primary limitations of CO_2 as a refrigerant:

- **Special components**. CO_2 systems operate at much higher pressures.

- **Energy efficiency that degrades at higher ambient temperatures**. The energy efficiency of CO_2 systems must be enhanced further—especially at higher ambient temperatures—to ensure that they can compete successfully against conventional systems in all climates.

- **Importance of system integration**. CO_2 systems can achieve significant improvements in energy efficiency if integrated and optimized.

- **Life-cycle cost performance (LCCP)**. Components for CO_2 systems are not yet mass produced, therefore, CO_2 equipment costs tend to be higher than those associated with conventional systems.

WATER HEATING

Hot water is used for a variety of daily functions in domestic applications including bathing, laundry, and dishwashing. Electricity is used to power electric resistance water heaters, electric heat pump water heaters, electrode boilers, pumps, and other devices such as microwave water heaters.

Commercial Building Electric Heat Pump Water Heaters

Heat pump water heaters operate on the same principle of heat pumps used for heating applications. They extract heat from a low-temperature source and transfer the heat to a high-temperature sink using electricity. With heat pump water heaters, the low-temperature source is typically the space surrounding the water heater and the high-temperature sink is the hot water tank.

The heat pump cycle employs an evaporator, compressor, condenser, expansion valve, and controls. In the heat pump water heater, heat from the surrounding air is used to evaporate a low-pressure liquid refrigerant in the evaporator. The vaporized refrigerant then enters the compressor, where its temperature and pressure are increased. Then the refrigerant enters the condenser where it transfers heat to the water and is condensed back to a liquid state. The pressure of the refrigerant reduces as it passes through an expansion valve.

Heat pump water heaters can come integrated with a water storage tank, or they can be added on to an existing storage tank. In addition, they can cool the space surrounding the water heater, or they can cool the outside air. They have the added benefit during the cooling season of cooling and dehumidifying the indoor air.

Heat pump water heaters are two to three times more energy efficient than electric resistance water heaters. Since the technology does rely on combustion processes to heat the water, it eliminate gaseous emissions (e.g., carbon monoxide) at the point of use.

Electric Boilers

In select applications and markets, the use of electric boilers is a viable alternative to fossil fuel-fired boilers. In part, the use of elec-

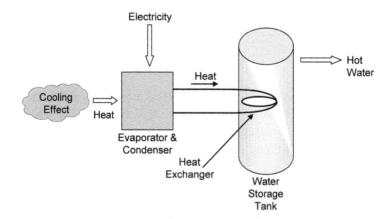

Figure 9-10. Heat Pump Water Heater Operation

tric boilers in conjunction with fossil fuel boilers can be a viable tool to commercial building operators interested in hedging against fossil fuel prices. Boilers are used to produce hot water or steam for a variety of commercial processes.

Electric boilers are available in sizes from 5 kW to 50 MW. There are two types of electric boilers: electric resistance and electrode boilers. Electric resistance boilers use resistance heating elements (under 600 V) to convert electric energy to heat energy. Electrode boilers are used in applications greater than 4 MW.

Electrode boilers usually operate at higher voltage (4 to 24 kV). There are two types: submersible electrode boilers and high-velocity jet electrode boilers. Submersible electrode boilers use immersed electrodes to conduct electricity through water. High-velocity jet electrode boilers rely on water jets striking an electrode plate. In these heaters, the water jet is the resistive element.

Electric boilers potentially have several advantages over fossil-fueled boilers. These may include:

- Energy Cost Savings—Where electric prices are low, electric boilers have discrete advantages in offering lower operating costs. Where demand-response or off-peak electric rates are in place, these electric boilers can be controlled for further operating cost savings.

- Higher Operating Efficiencies—Electric boilers have operating efficiencies of almost 100%, while fossil-fueled boilers have efficien-

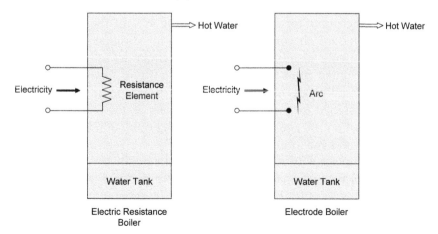

Figure 9-11. Illustration of Electric Boiler Types

cies between 60 and 85%.

- Smaller Footprint—Electric boilers have smaller footprints.

- Quieter Operation—Electric boilers are much quieter.

- Emission Reductions—Electric boilers reduce point source emissions and can reduce overall emissions if powered by low-carbon electric energy either on- and off-peak or solely off-peak.

Stand-alone heat pump water heaters use electric resistance elements as the backup heat source. Pre-heat systems work in conjunction with a conventional electric of fossil water heater.

The heat pump water heaters considered here are air-to-water units that are most often used in small- and medium-sized commercial applications. Larger water heating applications are addressed by heat recovery chillers and heat pumps.

Retrofit is plausible where there are existing fossil-fuel water heaters.

Heat pump water heaters are viable in most all buildings where water heating is used. All that is needed is a space of approximately 750 cubic feet for the water heater and to enable it to access ambient air. It is unlikely that this equipment would be used in conjunction with heat recovery chillers and heat pumps. Heat recovery chillers/heat pumps can provide water heating. Combining such heat recovery equipment with heat pump water heaters would be uneconomical.

Conditioning of outside air in a typical commercial air conditioning system accounts for approximately 30 to 40% of the overall cooling load. Standard air conditioning systems are designed with a fixed amount of outside air introduced, with the remainder of the supply air being re-circulated from the space. The quantity of ventilation air required is normally determined by the space design populace. Hence, if a room is designed for a maximum of 100 people, the ventilation provided is based on 100 people even though the population in the space is not normally at a maximum.

Use of inverter-driven fans in commercial buildings allows for demand control ventilation (DCV). DCV is a control strategy whereby the carbon dioxide level in the space is used as an indirect measure of the occupant density, and the ventilation air is modulated in response to the CO_2 levels. The use of a DCV control strategy can significantly reduce the amount of outside air required for a space, and consequently, the energy use of the air conditioning system.

EPRI conducted a scoping study in 2007, and based on this study and other energy simulation software, the annual kWh usage for DCV systems was estimated at around 60% of standard fixed ventilation systems. It is expected that the use of a DCV system could reduce total air conditioning energy usage by approximately 18 to 25%. This makes electric space conditioning far more competitive with gas-sired systems.

Microwave Water Heating

Microwave water heaters are tankless systems that produce hot water only when needed. Microwave water heaters consist of a closed stainless steel chamber with a flexible coil and a magnetron. When there is a demand for hot water, water flows into the coil and the magnetron bombards it with microwave energy. The microwave energy heats the water to the required temperature.

The microwave system can produce hot water in 20 to 30 seconds with continuous flow. Energy consumption is reduced compared to conventional fossil fuel-based water heaters. There is less pollution at the point of use since the technology does relay on combustion processes to heat the water, and it eliminates gaseous discharge.

FOOD SERVICE

There are key electric advantages in the commercial kitchen. They include:

- Improved Kitchen Comfort
  — Lower kitchen temperatures

- Reduced Operating Costs
  — Less ventilation required
  — Less maintenance

- Increased Productivity
  — Faster cooking
  — Unique, high-performance equipment

The bottom line is that electric cooking competes cost-wise with gas cooking even before the environment is considered. This is because electric cooking equipment is usually 2 to 3.5 times more energy-efficient than gas, overcoming the electric cost premium; maintenance and cleaning is often less with electric equipment, saving labor costs; and electricity's demand costs are less than expected.

Some high-performance food service equipment is only electric such as microwave ovens, induction cooking, and light-wave ovens. Microwave ovens cook food quickly; microwaves penetrate food to aid in cooking; equipment temperature is lower; and surfaces don't brown. In induction cooking, only the pan is heated thus less heat-transfer energy loss; there is less heat surrounding the cooking area; and there is accurate, fast control of heat to the pan. With light-wave ovens, foods cook fast, they brown the outside, and no preheating is required. Electric humidity-controlled cabinets are also available which keep food hot, moist, tasty, and permit food preparation flexibility.

Induction Cooking

There are a number of cooking appliances based on induction heating available. In induction cooking, electric energy is transferred to a ferrous metal surface through a magnetic field generated by a magnetic coil. The ferrous metal surface may be a pot or sauce pan on a range top, the heating tubes in a deep fat fryer, or the cooking surface of a griddle. A primary advantage of this technology is rapid, even heating. The induction range top has the added advantage that the range top surface is not made of ferrous material. Heating energy is transferred directly to the pot or pan. The range surface will warm up from conduction of heat from the cooking container but is usually only warm to the touch. Safety of kitchen staff is greatly improved due to reduction in potential burns.

Multi-Function Appliances

Multi-function appliances such as combination steamer/ovens and holding cabinets operate in several multi-function modes that were originally performed by separate pieces of equipment. If floor space is limited, combination steamer/ovens may replace stand-alone steamers and ovens. Key factors to evaluate are the frequency of simultaneous use of the steamers and ovens, the volume of food to be cooked, and the preparation schedule. In many cases, the combination units may be used as holding cabinets to keep prepared food warm prior to serving. Due to the superior moisture control in these units, better food quality is maintained longer.

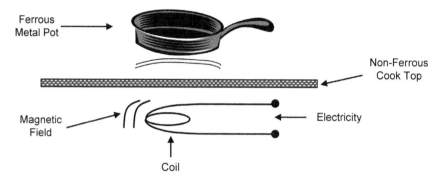

Figure 9-12. Illustration of an Induction Cook Top

Tilting skillets with lids are another innovative combination appliance. It is an insulated tilting skillet that can be used as a griddle, a steamer, and an oven, in addition to its traditional mode as a tilting skillet. The increased number of operating modes for this appliance are due to the use of an insulated compartment lid in the steamer and over modes.

Specialized Appliances
Cooking with Light

The Quadlux Corporation has a number of patents on its Flash-Bake® oven. This appliance is unusual in that it cooks with high-intensity light in the visible and infrared spectrums. It is particularly suited for pre-processed foods such as pizza and chicken fillets. It may be useful as an added appliance for multiple-station cafeterias or kiosks that will serve custom-prepared pizza or other quick-serve entrees.

High-Speed Convection Fryers

Another innovative appliance offered by a number of companies is a fryer that is essentially a convection oven using very high-speed hot air circulation to cook pre-processed frozen French fries. Commercially-prepared frozen French fries are partially cooked in frying oil and then quick-frozen. Convection fryers take advantage of the fact that enough oil is already in the product to produce a reduced-fat French fry. This appliance may be popular with food service operators that offer reduced-fat menus and want a quick-serve option for French fries.

Electric Fryer

Electric fryers are two to three times more efficient than conventional gas fryers at peak production rates and up to four times as efficient at lower (more common) production rates. This high energy efficiency often compensates for the higher price of electricity. In addition, the technology can help to extend fat life, thereby reducing fat costs.

Other Electric Cooking Equipment

Other electric cooking appliances now offer a variety of benefits. The advanced/innovative electric equipment includes the following:

- Induction fryers
- Solid-state fryers
- Convection/microwave ovens
- Air impingement/microwave ovens
- Convection/steam/microwave ovens
- Skittles
- Rofry (oil-free fryers)
- Electric rotisseries
- Electric conveyor broilers
- Blast chillers

LAUNDRY

Ozonated Laundering

Ozone (O$_3$) is a highly reactive oxidant that has been used for years to purify, disinfect, and deodorize water. Since 1990, ozone laundry systems have been installed in commercial laundries for washing fabric. Ozone has proven to quickly and effectively help break down soil, thereby reducing the amount of detergent and wash/rinse time needed. According to one explanation, when the electron-rich organics and hydrocarbons of laundry soils and stains are exposed to the electron-deficient ozone, an oxidizing reaction takes place, releasing the third oxygen atom from ozone. The highly electronegative oxygen atom then breaks many of the chemical bonds in the soils, fragmenting the molecules, making them easier to remove from the fabric. As a result, less water, less chemicals and detergents, less energy, and less wash time are required. In addition, the effluent wastewater's biological oxygen demand (BOD) and chemical oxygen demands (COD) are reduced by up to 50%, because the ozone dissipates into harmless oxygen.

Because ozone is highly reactive, it must be generated on-site as needed. Ozone can be generated by corona discharge or by ultraviolet (UV) excitation. In the corona discharge method, air or oxygen is passed between two electrodes, and a corona discharge is generated by applying high voltage to the electrodes. In the UV excitation method, ozone is generated photochemically by passing air through a path irradiated with UV light. In both methods, a portion of the air dissociates and recombines to form ozone. The most commonly used ozone generation method is corona discharge. The UV excitation method has limited use because of its high energy requirements (20 kWh per pound of ozone produced) and relatively low ozone production rate.

In an open-loop system, the soil-laden water is treated as effluent wastewater and discharged to the local treatment works. In a closed-loop system, the wastewater is drained from the washer, pumped through a screen or coarse bag filter, and then through an automatic backwashing sand filter—to remove all particles larger than 20 microns in diameter. The water then flows into a storage tank, where more ozone gas is injected and more microscopic bubbles are created. As the bubbles rise to the top of the tank, they oxidize odor molecules, viruses, and bacteria, and carry with them smaller remaining particles, such as oil and grease molecules. In the final step, the water is filtered a third time to remove particles larger than 5 microns and cleansed again with ozone. The recycled water is then ready for reuse in the washing machine.

OTHER COMMERCIAL APPLICATIONS

Ozonation of Swimming Pool Water

Viruses and bacteria survive in well-chlorinated pools and spas, mainly because of the buildup of total dissolved solids (such as body oils, suntan lotion, and cosmetics). As total dissolved solids build up, a buffer solution is produced, reducing the efficiency of chlorine as a sanitizer. Thus, viruses and bacteria—including pseudomonas—survive, even when the level of free chlorine residual exceeds most health department standards. Another disadvantage of chlorine is that it reacts with organic and inorganic contaminants in pools and spas to form toxic trihalomethane, a carcinogen. For these reasons, many lodging facilities are exploring alternative treatments.

Ozonation is a viable option; it reduces the need for chlorine by 70 to 97%. Ozone has one of the highest oxidation potentials, surpassed only by fluorine (F2) and the hydroxyl free radical (OH) in reactivity. In water treatment, ozone is typically the strongest oxidant used. It is about 3000 times more effective than chlorine in controlling viruses and bacteria. Ozone also destroys mold, fungi, spores, cysts, and algae, leaving no residue, and breaks down into harmless oxygen.

Because ozone is highly reactive, it must be generated on-site as needed. Ozone can be generated by corona discharge or by ultraviolet (UV) excitation. In the corona discharge method, air or oxygen is passed between two electrodes and a corona discharge is generated by applying high voltage to the electrodes. In the UV excitation method, ozone is generated photochemically by passing air through a path irradiated with UV light. In both methods, a portion of the air dissociates and re-combines to form ozone. The most commonly used ozone generation method is corona discharge. The UV excitation method has limited use because of its high energy requirements (20 kWh per pound of ozone produced) and relatively low ozone production rate.

Once ozone has been generated, it is then injected into the return line of the pool. When it comes in contact with the water, ozone destroys bacteria and viruses by bursting the cellular membrane and scattering the bacterium cytoplasm. This process takes about two to three seconds; with chlorine, the same process takes approximately 30 to 60 minutes.

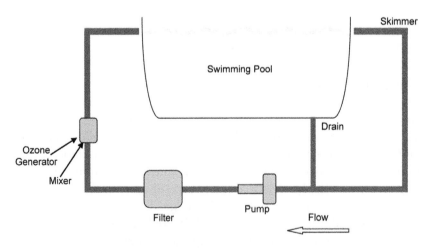

Figure 9-13. Illustration of Swimming Pool Ozonation

Reverse Osmosis of Drinking Water

Reverse osmosis is a membrane-separation technique that utilizes permeable membranes to filter selected components from a liquid. The membrane selects molecules on the basis of shape and size. Hotels could potentially use this process to reduce the levels of heavy metals, volatile organic compounds (VOCs), and chlorine in the drinking water.

Reverse osmosis is a very refined filtration method and utilizes a membrane with smaller pores than are used with microfiltration or ultrafiltration, two other methods of membrane separation. This system allows compounds smaller than 5 to 20 angstroms to pass through, while retaining larger compounds.

In a reverse osmosis system, water is circulated under pressure, in contact with a specially constructed polymeric film. Some dissolved matter passes through while other contaminants, such as heavy metals and chlorine, do not. The systems are modular, each designed as a self-contained pressure vessel containing the membranes and fluid distribution system. The systems typically operate at pressures of 200 to 1500 psi.

Four basic module designs or configurations exist:

- Tubular: The least susceptible to plugging, but the most expensive.

- Flat plate: Compact and less costly, with greater maintenance requirements.

- Spiral wound: Compact, low capital cost per unit, but requires more pre-filtration and makes leak detection more difficult.

- Hollow fiber: Relatively low capital cost, but requires pre-filtration and has limited operating pressures.

In all these configurations, solids build up on the membrane, usually because only hydrogen-bonding substances (e.g., water, ammonia) are allowed to pass through the membrane. The remedy for this membrane fouling involves either rinsing or back-flushing the membrane.

In rinsing, the membrane is flushed with feedwater at reduced pressures and increased velocity (two or three times normal). The turbulent action of the fluid loosens the particles and carries them away. Additives such as dilute hydrochloric acid, citric acid, dilute caustic soda, sodium hypochlorite, or detergent can assist in loosening the particles.

In back-flushing, the flow of the permeate (high-quality water) is reversed through the membrane. This process loosens and lifts particles from the membrane and washes them away with the concentrate (rejected particle stream). This ensures high filtration rates over long periods of time.

Infrared Sterilization of Medical Waste

The EPA estimates that seven million tons per year of medical waste is produced by health care facilities, of which a substantial portion is classified as infectious waste. Approximately one-quarter of all infectious waste is generated by non-hospital facilities such as clinics and laboratories. Disposing of infectious waste is becoming increasingly difficult due to tougher federal and state emission standards affecting incinerators and shrinking landfill space. For these reasons, medical practitioners and clinics are seeking new technologies that are cost-effective and environmentally safe. One new technology that has emerged is infrared (IR) sterilization.

Infrared sterilization uses IR rays to treat infectious waste quickly and effectively. The rays focus on and are absorbed by the "target" wastes. This target-specific sterilization transfers heat directly to the waste load, instead of to the air. The direct heating process is superior to convection heating since "cold spots"—areas where microorganisms can survive—are prevented. No chemicals, steam, or effluent are generated in the process. The infectious waste is reduced by 90% in volume, and the sterilized residue is suitable for municipal landfill disposal.

IR sterilization systems use re-circulating convection heat, conduction, and far IR rays to process infectious wastes. These units typically have three components: a computer system, a treatment chamber, and a volume reduction chamber. An operator loads infectious waste onto a revolving tray carrier attached to the access door. Upon closing the door and pushing the start button, the computer system takes control. The waste is heated to 425°F for 25 minutes and is then transferred to the volume reduction chamber automatically via a robotic mechanism. Once this occurs, the treatment chamber is ready for another load. Inside the reduction chamber, the sterilized waste is allowed to cool, then is crushed into an unrecognizable residue suitable for landfill disposal. The system is designed to treat a variety of wastes including needles, glass tubes, blood, IV bags, and disposable gloves. These units cannot, however, handle anatomical, radiological, or chemotherapeutic wastes; flammable hydrocarbon materials; and aerosol cans.

Low-Volume Infectious Waste Treatment

According to the U.S. Environmental Protection Agency (EPA), non-hospital health care facilities produce several hundred thousand tons of infectious waste annually in the United States. On average, non-hospital medical facilities produce 1000 pounds of infectious waste per month. Some facilities, such as doctors' offices, dental offices, nursing homes, and funeral homes, produce less than 50 pounds of infectious waste per month. Since these facilities individually generate such low volumes of infectious waste, little attention has been paid to developing treatment technologies for this category of infectious waste generators. Yet these facilities face the same medical waste disposal problems as hospitals, such as liability, potential threat to occupational safety and health, increasingly stringent regulations, and the rising cost of traditional methods of treatment and disposal.

However, a number of alternative infectious waste treatment technologies are available for low-volume generators. These technologies can be grouped into four categories, based on the type of treatment processes utilized: chemical-mechanical, electro-chemical, resistance sterilization, and thermal-mechanical.

Chemical-Mechanical

This type of process utilizes chemical and mechanical methods to destroy and decontaminate biohazardous waste such as syringes, needles, glassware, laboratory waste, blood and other body fluids, specimens, cultures, and other contaminants. It cannot, however, treat chemotherapeutic and pathology wastes or hazardous chemicals.

In this process, infectious waste is placed into a portable chamber at the point of generation. When the chamber if full, it can be transported to the processor. At this point, a specific dose of decontaminant—a peracetic acid formulation—is added to the waste. The waste is then mechanically ground and chemically decontaminated. At the end of a 10-minute cycle, the liquid by-products—vinegar and hydrogen peroxide—are separated from the solid waste and discharged into the sewer. The solid waste can then be sent to the landfill with the regular municipal solid waste. This system can treat up to 8 pounds of waste per 10-minute cycle.

Electro-Chemical

Like the chemical-mechanical process, this method can decontaminate medical waste including sharps, blood and other body fluids, cultures, dialysis waste, lab waste, disposable gowns, gloves, and masks, but cannot treat chemotherapeutic and pathology wastes or hazardous chemicals.

In this type of system, up to 20 pounds of infectious waste is placed in a removable basket and immersed in a sodium chloride (NaCl) and water mixture. Electrodes at the bottom of the container liberate chlorine, ozone, and their respective hydroxyl radicals, which decontaminate the waste. This process destroys all pathogens and spores in approximately five minutes, thus rendering the waste safe and legal for disposal in a municipal landfill.

Resistance Sterilization

Heat and time, when combined, are an effective method of sterilization. Laboratory tests show that no microorganism can survive temperatures about 356°F for more than a fraction of a minute. Based on this principle, a resistance sterilization unit has been designed to treat sharps (syringes, needles, and scalpels).

In the resistance sterilization process, infectious sharps are collected in a plastic container near the point of contamination. When full, the container and a temperature-indexed plastic disk are placed in the processor. Thermal energy then melts the container and specially-designed disk, creating a molten liquid that flows through the waste. The plastic disk is certified not to melt below 375°F. The melting of the disk is, therefore, evidence that temperature and time requirements have been met and serves as a biophysical indication of sterilization. The sterilized material cools into a solid plastic block that can be discarded as municipal solid waste.

Thermal-Mechanical

This treatment system uses heat and mechanical action to render infectious sharps and rigid plastic waste sterile and unrecognizable and, therefore, suitable for municipal landfill disposal. Two systems are currently available: a small unit that can treat 40 syringes an hour, and a larger unit that can treat up to 16 gallons of sharps and rigid plastic every 105 minutes. In both units, the waste is collected in a specially-designed container at the point of generation. When the container is full, it is fed into a central processor. In the smaller system, the waste is ground and then heated for 30 minutes at 480°F, by transforming it into a non-infectious plug that is approximately 15% of its original volume and about 4 inches in diameter. In the larger unit, the infectious waste is heated at 535°F and converted into a non-infectious block that is approximately 20% of its original volume. An optional grinder can be used to reduce the waste to unrecognizable shavings.

MEDICAL ELECTRONICS

Applications for electricity in medicine have offered some of the greatest advances for improving the quality of life for every human being. In every aspect of modern medicine, electrically powered equipment diagnoses and treats illnesses; virtually every modern medical tool substantially enhances the profession's capability to save and give life. Today, medical technology employs electricity in a myriad of ways. For example, doctors use it to stimulate muscle tissue and nerves to relieve pain, reinstate hearing in the deaf, and maintain a regular heartbeat. In addition, a new and particularly exciting application involves using electricity to combat neuromuscular disorders that cause paralysis.

Researchers learn daily about the effects of small electric impulses on pain. Pain is an important indicator of where we hurt and what is wrong. Researchers are finding that electrical stimulation of muscle tissues and nerves causes the body to release naturally produced opiates—called endorphins—which either suppress pain or block the transmission of pain signals to the brain. This naturally induced "pain killer" often relieves pain more efficiently than pharmaceutical drugs, many of which simply mask it.

Researchers have devised electrical systems that stimulate hearing in some profoundly deaf people. These systems consist of a microphone and transmitter located outside of the person, an implanted receiver, and an electrode located near the inner ear. The external system picks up sound, transmits it inward, and the receiver converts the signals for the inner ear to decipher.

Millions of people rely on pacemakers to regulate their heartbeats. This technology, one of the most commonly cited examples of electricity's ability to improve one's quality of life, has been in use since 1958. New generations of this technology can pace both the heart's right ventricle and atrium, and allow telemetry of the heart's activity.

Another recent, exciting application of electricity in medicine involves neuromuscular stimulation (NMS). NMS involves electrically stimulating a muscle so strongly that it contracts. Researchers are finding that this action often produces a powerful motor reaction which restores some level of function to dysfunctional limbs. Basically, NMS may enable paraplegics and quadriplegics to walk! Thus, electricity may successfully treat several neuromuscular disorders; this research is rapidly evolving and early results prove quite promising.

References

"Variable Refrigerant Flow Air Conditioners and Heat Pumps for Commercial Buildings," EPRI, Palo Alto, CA: 2008. 1016258.

"Residential Heat Pump Water Heaters," EPRI, Palo Alto, CA: 2007. 1015428.

"Beneficial Electrification: An Assessment of Technical Potential," EPRI, Palo Alto, CA: 1992. CU 7441.

"Program on Technology Innovation: Advanced Technologies for Energy Efficiency in Residential and Commercial Buildings," EPRI, Palo Alto, CA: 2008. 1016875.

"Medical Clinics: A Small-Business Guide," EPRI, Palo Alto, CA: 1996. 106676-V4.

"Lodging: A Small-Business Guide," EPRI, Palo Alto, CA: 1996. 106676-V3.

Chapter 10

Beneficial Residential Building Uses of Electricity

As in other sectors, there are ample opportunities for the increased penetration of beneficial new uses of electric devices and appliances in the residential sector. These are predominantly in residential heating, cooling, water heating, cooking, and laundry. This chapter explores several of those applications.

SPACE CONDITIONING (HEATING AND COOLING)

There are an increasing number of technologies which can displace inefficient fossil-fueled appliances to provide heating and cooling.

Air-Source Heat Pumps

Air-source heat pumps provide important options for residential space heating and cooling applications. Residential heat pumps consist of single-phase equipment under 65,000 Btu/hr.

An air-source heat pump is a device for absorbing heat from the air and transferring that heat to a conditioned space or vice-versa. When operating in the heating mode, the air-source heat pump absorbs heat from the ambient air and raises it to a higher temperature; delivering it to the space. In the cooling mode, heat is removed from the space, elevated to a higher temperature and rejected to the ambient air. The majority of air-source heat pumps used for space conditioning in the United States deliver the heating or cooling to the space with air distribution, either through duct work or with air flowing over the refrigerant coil (EPRI 1016075).

Most heat pumps use the vapor compression cycle, sometimes called the (reverse) Rankine cycle. Vapor compression cycle heat pumps use a volatile working fluid, a refrigerant, and the processes of vapor compression, condensation, pressure reduction, and evaporation. Refer to Chapter 9 for a more thorough description.

The major components of a vapor compression cycle heat pump are the compressor, heat exchangers (a condenser and an evaporator), and expansion device. Other devices include fans and blowers, suction line accumulators, a liquid line filter dryer, a reversing valve, supplementary heaters and controls.

Types of Air-Source Heat Pumps

There are several types of air-source heat pumps including single package, split, dual-fuel, booster, mini-split, multi-split, variable refrigerant flow systems, room heat pumps, and packaged terminal heat pumps.

Unitary Air Source Heat Pumps

Unitary heat pumps are factory-made assemblies that include an evaporator, condenser, and compressor. A heat pump with one factory-built assembly is called a single-package system, and a heat pump with more than one factory-built assembly (indoor and outdoor units) is commonly called a split system.

Single-package heat pumps are configured as self-contained, factory-assembled modules that contain all the components of the refrigeration system: compressor, indoor and outdoor coils, fans and controls, and supplemental heating.

The split system is the most common type of heat pump sold in the United States, consisting of two or more factory-built modules to be assembled in the field: An outdoor unit containing the outdoor heat exchanger (coil), fan, fan motor, compressor; a reversing valve, an indoor unit or air handler with a blower, blower motor, indoor heat exchanger; and electric resistance heating elements. In some models, the compressor is located in a separate cabinet indoors. The indoor and outdoor units are connected by insulated refrigerant tubing that is put in place when the heat pump is installed.

Ductless Mini-Split, Multi-Split and
Variable Refrigerant Flow Heat Pump Systems

Ductless heat pumps (DHPs), also know as mini- and multi-split heat pumps are non-ducted, split-system heat pumps that were developed in Japan in the 1950s as quieter, more efficient alternatives to window units.

Products have evolved from a single indoor unit operating off each outdoor unit to current variable refrigerant flow (VRF) configurations with refrigerant lines. A DHP mini-split product for residential application is shown in Figure 10-1 consisting of one outdoor unit and one or more indoor fan-coil units. Like a central (ducted) split-system heap pump, the outdoor unit can be installed on grade or on the roof, placed on the balcony of a high-rise apartment building, or hung.

VRF systems connect the units in a manner that minimizes refrigerant piping and permits heat recovery between units.

Dual Fuel Heat Pumps

Dual fuel heat pumps are available in both single-package and split-system configurations consisting of a heat pump and a forced air furnace (most commonly a natural gas-fired furnace), and as add-on equipment versions where a heat pump is added to a furnace to provide both electric heating and cooling replacing the air conditioning unit previously used just for cooling.

Figure 10-1. Ductless Mini-Split Heat Pump System

Booster Systems

In recent years, products have entered the market using two compressors in series with an economizer-type liquid subcooler designed to provide increased mass flow, capacity, and efficiency at low outdoor temperature.

Room Heat Pumps

Room heat pumps (also called reverse-cycle room air conditioners) are designed to heat and cool single zones and are best suited to applications where ducts are not available for providing space conditioning or where single zones need to be conditioned separately. Room heat pumps are available in nominal cooling capacities of about 7000 to 17,000 Btu/h.

In a typical room heat pump, both indoor and outdoor sections are encased in a single cabinet that is installed in a window or through the wall. The outdoor section usually has side and back louvers that serve as the air intake and exhaust. Therefore, a part of the unit extends out from the side of the building to allow outside air to be drawn in and exhausted out after passage over the outdoor coil. The indoor section serves as both a return air intake and a conditioned air supply.

Closed-loop systems use an underground network of sealed, high-strength piping, which acts as the earth-coupled heat exchanger. The most commonly used closed-loop piping material is high-density polyethylene (HDPE). However, new polymers such as PEX are being introduced and accepted by the industry. The ground loop piping is filled with a working fluid that is continuously re-circulated without ever directly contacting the soil or water in which the loop is buried or immersed. Once filled with fluid and purged of air, nothing enters or leaves the closed loop. This eliminates the potential shortcomings of water quality and availability associated with open-loop systems.

Geothermal Heat Pumps

Much like air-source heat pumps, geothermal heat pump (GHP) systems are essentially air conditioners that can also run in reverse to provide heat in the winter. The primary difference is that they rely on the nearly constant temperature of the earth for heat transfer instead of the fluctuating temperatures of the outside air; consequently, they are far more efficient than most alternative systems.

GHP systems utilize a refrigerant to help transfer (or pump) heat into and out of a home or building. The refrigerant allows the GHP system to take advantage of two primary principles of heat transfer:

1. Heat energy always flows from areas of higher temperature to areas of lower temperature.

2. The greater the difference in temperature between two adjacent areas, the higher the rate of heat transfer between them.

Because air-source heat pumps extract heat from the outside in the winter and reject heat outside in the summer, their efficiency is lowest when the need is highest. As the temperature difference between the air and the refrigerant decreases, the heat transfer rates drop, and consequently, the systems efficiency drops as well. By using the relatively constant temperature of the earth as a heat source in winter and a heat sink in summer, instead of the varying outside air, GHP systems overcome the limitations of air-source heat pumps thereby allowing for much higher system efficiencies. With ground temperatures almost constant year round, between 45° to 70°F across the U.S., GHPs can leverage the higher temperature differences between the refrigerant and the ground, therefore achieving much higher efficiencies than standard heat pumps.

To extract and reject heat from the ground, GHP systems use a ground loop (heat exchanger). The ground loop can either be open-loop, closed-loop, or direct-exchange (DX).

The biggest benefit of GHPs is that they use 25 to 50% less electricity than conventional heating or cooling systems. This translates into a GHP using one unit of electricity to move three units of heat from the earth. According to the EPA, geothermal heat pumps can reduce energy consumption—and consequently emissions—up to 44% compared to air-source heat pumps, and up to 72% compared to electric resistance heating with standard air conditioning equipment. GHPs also improve humidity control by maintaining about 50% relative indoor humidity, making GHPs very effecting in humid areas.

GHP technology has several benefits including:

• **Low Operating Cost**—The efficiency of the heat pumps operating under moderate loop temperatures provides the basis for high efficiency and low operating cost.

- **Simplicity**—The distributed nature of the system makes it easy to understand. A heat pump located at each space will provide independent heating and cooling.

- **Low Maintenance**—The heat pump itself is a packaged unit no more complex than typical residential air conditioning equipment.

- **No Supplemental Heat Required**- Heat pumps can meet all of the space loads including ventilation loads.

- **Low-Cost Integrated Water Heating**—Heat pumps can be dedicated to meet hot water loads.

- **No Required Exposed Outdoor Equipment**—The ground heat exchanger is buried, and the heat pumps are located inside the building.

- **Low Environmental Impact**—No fossil fuels need to be consumed on site. In addition, since GHP units are more efficient than other types of heating technology, overall CO$_2$ impacts are minimal.

- **Level Seasonal Electric Demand**—With winter heat pump operation displacing fossil fuel use, and summer heat pump operation occurring at moderate, more efficient loop temperatures, the electric demand is more consistent throughout the year.

- **Longer Life Expectancy**—The appropriate service life value for ground source heat pump technology is 20 years or more.

- **Improved Utility Asset Utilization**—The use of GHP results in increased electric load, improved asset utilization through higher annual load factors, and slower peak load growth.

A hybrid geothermal system uses geothermal loop along with elements of "conventional" water-loop systems to reduce overall system costs which in turn provides benefits of lower peak demands, greater load factors, and carbon footprint reduction to utilities. In some applications, the seasonal heating loads can be larger than the cooling loads where loop size may be governed by the heating loads. In such cases, a hybrid solution would reduce the ground loop size by providing supplemental heating to the loop to avoid long-term buildup of excess cooling in the ground. Such heat can be provided from a conventional boiler, a solar water heater, or heat from any waste heat stream.

WATER HEATING

Residential Heat Pump Water Heaters

Heat pump water heaters (HPWH) are significantly more energy efficient than electric resistance water heaters and can result in lower annual water heating bills for the consumer, as well as reductions in greenhouse gas emissions.

Heat pump water heaters, which use electricity to power a vapor-compression cycle to draw heat from the surrounding environment, can heat water more efficiently for the end user than conventional water heaters (both natural gas and resistant element electric). Such devices offer consumers a more cost-effective and energy-efficient method of electrically heating water. The potential savings in terms of carbon emissions at the power plant are also significant. Replacing 1.5 million electric resistance heaters with heap pump water heaters would reduce carbon emissions by an amount roughly equivalent to the annual carbon emissions produced by a 250 MW coal power plant.

Operation of heat pump water heaters are based on the same thermodynamic cycle found in the common household refrigerator or air conditioner, the vapor compression cycle. A fluid refrigerant is circulated through a loop, in which it successively undergoes expansion, evaporation, compression, and condensation.

In most residential heat pump water heaters, the evaporator is placed in contact with the surrounding air, allowing the construction of water heaters that fit the same footprint as conventional water heaters. Such water heaters are commonly called "drop-in" designs. There are two predominant types of drop-in heat pump water heaters: Integral heat pump water heaters which replace existing water heaters, and remote heat pump water heaters which add a heat pump unit to the tank of an existing water heater.

The use of the thermodynamic cycle allows heat pump water heaters to have energy efficiencies much higher than conventional water heaters. In conventional water heaters, some of the energy input is lost to the surroundings, so that the efficiency is invariably less than 100%. In heat pump water heaters, however, the system actually draws heat from the surroundings, resulting in a heat energy output greater than the electrical energy input. This makes it seem as though the efficiency of the system is actually greater than 100%. To prevent confusion with thermodynamic efficiency (which can never be greater than 100%), the measurement of efficiency of heat pump water heaters is called the co-efficient of performance (COP), which is defined as the useful energy output (in the form of the thermal energy used to heat water) divided by the energy input (typically electricity or a combustible fuel).

The COP of heat pump water heaters typically ranges between 2.0 and 4.0, meaning that for every unit of electrical energy used for operation, two to four units of energy are transferred to the water. Japanese Eco Cute heat pump water heaters (described below) have demonstrated a COP as high as 4.8 (EPRI 1015428).

RESIDENTIAL APPLIANCES

The major traditional residential appliances are those referred to as "white goods":

- Refrigerators and freezers
- Cooking equipment
- Clothes washers and dryers
- Dishwashers

Residential appliances account for about 20% of an average U.S. household's electricity use, excluding energy use by electric water heaters that serve the clothes washer and dishwasher (about 45% of water heaters in the U.S. are electric). A breakdown of average electricity consumption by end use is provided in Figure 10-1.

Induction Cooktops

Induction cooking offers significant improvements in energy efficiency compared to other cooktops. They achieve efficiencies in the range of 80 to 95%, compared to efficiency of conventional smooth-top electric cooktops of 50 to 65%. Also, since heat is generated in the food container rather than at the burner, this form of cooking can also reduce the amount of heat being released into the kitchen. It also is an attractive option for customers with all-electric service—not only because it is energy efficient, but because it compares favorably with gas cooktops in terms of features valued by cooks, such as responsiveness, precise temperature control, and safety.

Electric cooktops can be divided into five types:

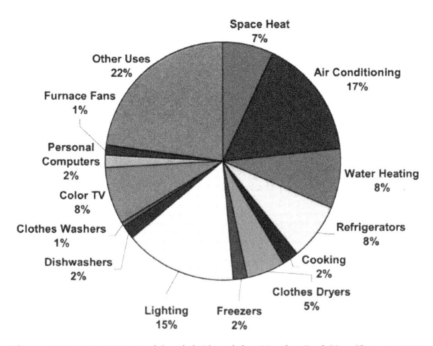

Figure 10-2. 2008 U.S. Residential Electricity Use by End Use (Source: 2008 Annual Energy Outlook)

- Electric coil
- Solid disk
- Halogen
- Radiant
- Induction

Electric-coil cooktops have the familiar coiled elements that are heated by electric resistance. Solid-disk designs are similar to exposed electric-coil units, except the coiled electric elements are configured in a solid design for easier cleaning.

Halogen and radiant cooktops are frequently referred to as "smooth-top" designs. In both halogen and radiant technologies, the heating elements are placed beneath a smooth ceramic glass surface which facilitates easy cleaning. Halogen systems use a quartz halogen lamp, and radiant designs use an electric coil. In both halogen and radiant units, the heating elements radiate energy through the ceramic glass to the cookware.

Induction systems, which are typically designed as smooth tops, transfer energy to cookware through a magnetic field.

The relative ranking of efficiency for these technologies is as follows (from lowest to the highest): solid disk (lowest), electric coil, radiant smooth top, halogen smooth top, and induction (highest).

If industry efficiency estimates are used, rather than DOE efficiency numbers, the savings are even larger. Using 40% for gas, 55% for electric coil, 60% for electric smooth top, and 90% for induction, the energy savings for induction becomes 56% compared to gas, 39% compared to electric coil, and 33% compared to electric smooth top (see Table 10-1).

Heat Pump Clothes Dryers

Heat pump dryers rely on the vapor compression cycle for extracting heat, similar to a window air conditioner operating in reverse. Hot moist air is channeled through the heat pump, where moisture is condensed and sent to a drain, and then the air is reheated by the heat pump, and the dehumidified warm air is re-circulated through the drum. Warm air is not exhausted, it is conserved within the machine, one of the reasons these dryers can use up to half of the electricity consumed by conventional dryers.

Heat pump dryers are available in Europe and Asia. A heat pump unit offers the greatest opportunity for energy savings in dryers since they can halve the amount of energy used. A DOE assessment done by TIAX in 2005 of estimated energy savings achieved by commercially available heat pump dryers in 2005 was 30 to 50% (DOE, 2009).

Microwave Clothes Dryers

Microwaves can dry clothing as they penetrate easily to the moisture in the interior of fabrics and vaporize it. Use of microwaves for clothes dryers have been investigated by appliance manufacturers, as well as by electric utilities.

EPRI developed a microwave dryer test unit that combined a conventional electric resistance heating element with magnetrons for drying. The unit could operate in three modes: cool drying (microwave power and ambient air), fast drying (microwave power and resistance element heating), and high-efficiency drying (microwave power and waste heat recovery). Which mode of operation was used had an impact on efficiency, with the efficiency of the test units ranging from slightly

Table 10-1. Energy Comparison Based on Industry Efficiency Values (EPRI 1018980)

| Type of Cooktop | Output Energy (1) | Efficiency | | Input Energy (!) | Energy Reduction if Induction Used |
		Range	Used for Calculations		
Gas		30-45%	40%	2.50	56%
Electric Coil	1.0	45-60%	55%	1.82	39%
Electric Smooth Top (2)		50-65%	60%	1.67	33%
Induction		80-95%	90%	1.11	---

Notes:
(1) Output energy and input energy are expressed without dimensions in this table.
(2) Smooth top is presumed to have a halogen element.

worse than conventional dryers to 13 to 25% better. A residential clothes
dryer prototype was developed but was never commercialized, primarily because the improved efficiency attained would have had a cost that
resulted in a simple payback period exceeding the likely lifetime of the
dryer.

CONCLUSIONS

While the residential sector has the least array of individual applications for beneficial new uses—it is a most important sector.

References
"Residential Heat Pump Water Heaters," EPRI, Palo Alto, CA: 2007, TR-1015428.
"Air Source Heat Pumps for Residential and Light Commercial Space Conditioning Ap-
      plications, Technical Brief," EPRI, Palo Alto, CA: 2008, TR-1016075.
"Variable Refrigerant Flow Air Conditioners and Heat Pumps for Commercial Buildings,
      Technical Brief," EPRI, Palo Alto, CA: 2008, TR-1016258.
"Geothermal Heat Pumps Technical Update: Technology & Market Overview," EPRI,
      Palo Alto, CA: 2008.
"Hybrid Geothermal Heat Pump Systems," EPRI, Palo Alto, CA: 2009, TR-1017888.
"Residential Appliances: Energy Efficiency and Technology Trends," EPRI, Palo Alto,
      CA: 2009, TR-1018980.

Chapter 11

Enhancing
Energy Efficiency

Reducing CO_2 emissions in modern society will require three actions: making end-use applications of electricity as efficient as possible; converting the majority of fossil-fired end uses to electricity; and de-carbonizing the electricity sector. This chapter describes actions to make the end use of electricity as efficient as practical. Its principles are needed equally in both programs to increase efficiency and programs to convert fossil-fueled uses to electricity. Even when conversion of fossil-fueled end uses to electric end uses is undertaken—to succeed, the applications must be as efficient as practical.

The world is looking to energy efficiency to help meet the challenges of maintaining reliable and affordable electric service, wisely managing energy resources, and reducing carbon emissions. Many states have established legislation to mandate energy efficiency savings levels and regulatory mechanisms to allow utilities to make energy efficiency a sustainable business. Fundamental to such policies are estimates of the potential for energy efficiency grounded in technological expertise and tempered by economic and market realities.

According to the Energy Information Administration's 2008 Annual Energy Outlook (AEO 2008) Reference Case, annual electricity consumption for the U.S. in the residential, commercial, and industrial sectors is estimated at 3,717 TWh in 2008. The AEO 2008 Reference Case forecasts this consumption to increase by 26% to 4,696 TWh in 2030, an annualized growth rate from 2008 to 2030 of 1.07%.

The AEO 2008 Reference Case already accounts for market-driven efficiency improvements and the impacts of all currently legislated federal appliance standards and building codes .

Total U.S. spending by utilities on electric end-use energy efficiency programs and initiatives has doubled in two years from $2.2 billion in 2007 to $4.4 billion in 2009 (DOE, 2010). As shown in Table 11-1, electric energy efficiency savings, as reported by state, now range up to 1.3% savings per year.

Energy efficiency is also a priority elsewhere in the country. I fact, many states are enacting their own legislation setting targets. For example: under Maryland's EmPOWER initiative, the state will reduce energy consumption by 15% by 2015; Pennsylvania's Act 129 requires a 1% reduction in consumption by May 31, 2011 and a 3% reduction in consumption (as well as a 4.5% reduction in peak demand) by May 31, 2013; the Arizona Corporation Commission requires electric utilities to reduce the amount of power they sell by 22% by 2020; and New Mexico has a stated goal of 20% reduction by 2020 (Faruqui, 2010).

A recent EPRI study (EPRI 1016987) predicts that energy efficiency programs have the potential to reduce electricity consumption in 2030 by 398 to 544 billion kWh. This represents a range of achievable potential reduction in electricity consumption in 2030. Relative to the AEO 2008 Reference Case, which assumes a level of energy efficiency program impact, the EPRI study identifies between 236 and 382 billion kWh of additional saving potential from energy efficiency programs. Therefore, energy efficiency programs have the potential to reduce the annual growth rate in electricity consumption forecasted in AEO 2008 between 2008 and 2030 of 1.07% by 22% to 36%, to an annual growth rate of 0.83% to 0.68%.

Table 11-1. Top 10 Current Electric Energy Efficiency Savings by State (2007 data, ranked by total MWh savings)

Rank	State	Total Incremental Electricity Savings (MWh)	Savings as Percent of Electricity Sales
1	CA	3,393,016	1.3%
2	WA	635,062	0.7%
3	NY	540,612	0.4%
4	MA	489,622	0.9%
5	WI	467,725	0.7%
6	MN	463,543	0.7%
7	TX	457,808	0.1%
8	OR	437,494	0.9%
9	CT	371,899	1.1%
10	fl	348,208	0.2%

Source: ACEEE 2009 State Energy Efficiency Scorecard

The estimated levels of electricity savings are achievable through voluntary energy efficiency programs implemented by utilities or similar entities were included in the EPRI study. The analysis did not assume the enactment of new energy codes and efficiency standards beyond what is already in law. More progressive codes and standards would yield even greater levels of electricity savings than elucidated in the EPRI study.

The EPRI study used an analysis approach that is consistent with generally prescribed methods. The study applied two distinct approaches to estimate electric energy efficiency: one for residential and commercial buildings and another for industrial facilities. For the residential and commercial sectors, the study implemented a bottom-up approach for determining electric energy efficiency savings potential. The residential and commercial approach begins with a detailed equipment inventory (e.g., the number of refrigerators), the average unit energy consumption (per household or per square foot in the commercial sector), and the diversified load during the non-coincident summer peak. In each sector, annual energy use and peak demand are the product of the number of units and the unit consumption annually, and at peak. This process is repeated for all devices across vintages and sectors. AEO 2008 provided both the number of units and the unit consumption. For the industrial sector, the study applied a top-down approach in which the sector forecast is allocated to end uses and regions.

The savings potential of an individual energy-efficiency measure is a function of its unit energy savings relative to a baseline technology and its technical applicability, economic feasibility, the turnover rate of installed equipment, and market penetration. For a given end use, a baseline technology represents a discrete technology and is generally the most affordable and prevalent technology option in its end-use category.

For example, for residential central air conditioning (CAC), the baseline technology is a unit with a seasonal energy efficiency ratio (SEER) of 13. In the EPRI analysis, the baseline SEER 13 unit, along with more efficient and expensive, technology options (e.g., SEER 14, SEER 15, SEER 17, ductless inverter-driven mini-split heat pumps, etc.) are applicable in existing homes as replacements for CACs that have reached the end of their expected useful life. They are also applicable to new homes that are being built with CAS.

The EPRI study utilized a modeling tool for forecasting energy use and energy efficiency savings. The modeling approach is consistent with EPRI's end-use econometric forecasting models including Residential End-Use Econometric Planning System (REEPS) and the Commercial End-Use Planning System (COMMEND), which are detailed microeconomic models that forecast energy and peak demand at the sector, segment, and end-use levels. The modeling tool used in this study represents a simplification of these legacy EPRI models customized for the analytical task of estimating energy efficiency potential. The study incorporates a comprehensive technology database that includes the latest findings from EPRI energy efficiency research. Energy efficiency savings potentials are developed using a bottom-up approach, aggregating the impact of discrete technology options within end uses across sectors and regions. This approach follows industry best practices and has been applied successfully in numerous forecasting and potential studies for utilities.

The EPRI study applied the condition that new equipment does not replace existing equipment instantaneously or prematurely, but rather is "phased in" over time as existing equipment reaches the end of its useful life.

The study focused on the technical potential which represents the savings due to energy efficiency and demand response programs that would result if all homes and businesses adopted the most efficient, commercially available technologies and measures, regardless of cost. Technical potential provides the broadest and largest definition of savings since it quantifies the savings that would result if all current equipment, processes, and practices in all sectors of the market were replaced at the end of their useful lives by the most efficient available options. Technical potential does not take into account the cost-effectiveness of the measures or the rate of market acceptance of those measures (i.e., 100% customer acceptance assumed).

Using a residential central air conditioning example, technical potential assumes that each year every home with a residential central AC unit that has reached the end of its useful life purchases and installs the most efficient technology as a replacement (i.e., ductless inverter-driven mini-split heat pumps).

In the EPRI study, maximum achievable potential (MAP) was used to take into account those barriers that limit customer participation under a scenario of perfect information and utility programs. MAP involves incentives that represent 100% of the incremental cost of energy efficient measures about baseline measures, combined with high administrative and marketing costs. These barriers could reflect customers' resistance to doing more than the absolute minimum required or a dislike of the technology options. For example, some customers might choose not to buy compact fluorescent lamps (CFLs) because they don't like the color or don't believe they work as well as incandescent lamps. When considering the purchase of major appliances, many customers consider price, aesthetics, and functional attributes before turning to energy efficiency and operations costs. Similarly, even through a financial incentive such as a rebate afforded by a program would bring the up-front cost of an energy-efficient product at parity with a standard product, some segment of customers would not be willing to go through the perceived hassle of a rebate application. This despite the clear economic benefits that would accrue from the monthly bill savings that result from a more efficient device. MAP is estimated by applying market acceptance rates (MARs) to the economic potential savings from each measure.

AEO Baseline forecast in 2030 is based on a maximum achievable potential of 544 TWh, or an 11% reduction in projected consumption. Relative to the AEO 2008 Reference Case, in 2030, this means a maximum achievable potential of 382 TWh of additional energy efficiency savings, or an 8% reduction in projected consumption.

The following is an example of the residential air conditioner to illustrate the technical potential. Central air conditioning (CAC) systems in existing homes are replaced, upon reaching the end of their useful lives, with the highest SEER-level equipment available within reasonable cost. In new homes, the highest SEER level available in each year is installed. In 2010, this is the SEER 20 air conditioner or the ductless (mini-split) heat pump with variable-speed operation.

Table 11-2 summarizes the energy-efficiency measures which may results in electric savings.

OTHER ESTIMATES OF THE POTENTIAL
FOR ENERGY EFFICIENCY

Since the mid-1980s, there have been a variety of studies conducted for the purpose of assessing the true potential for energy efficiency. These studies often vary significantly in their results. The primary reason for the difference in their results stems from the range of assumptions used.

Table 11-2. The Most Effective Energy Efficient Measures

Residential Sector Measures	Commercial Sector Measures
Efficient air conditioning (central, room, heat pump)	Efficient cooling equipment (chillers, central AC)
Efficient space heating (heat pumps)	Efficient space heating equipment (heat pumps)
Efficient water heating (e.g., heat pump water heaters and solar water heating)	Efficient water heating equipment (heat pumps
Efficient appliances (refrigerators, freezers, dishwashers, clothes washers, clothes dryers)	Efficient refrigeration equipment and controls (e.g., efficient compressors, floating head pressure controls, anti-sweat heater controls, etc.)
Efficient lighting (CFL, LED, linear fluorescent)	Efficient lights (interior and exterior, LED exit signs, task lighting)
Efficient power supplies for information technology and consumer electronic appliances	Lighting controls (occupancy sensors, daylighting, etc.)
Air conditioning maintenance	Efficient power supplies for information technology and electronic office equipment

Table 11-2 (*Cont'd*). The Most Effective Energy Efficient Measures

Residential Sector Measures	Commercial Sector Measures
Heat pump maintenance	Water temperature reset
Duct repair and insulation	Efficient ventilation (air handling and pumps, variable air volume)
Infiltration control	Economizers and energy management systems (EMS)
Whole-house and ceiling fans	Programmable thermostats
Reflective roof, storm doors, external shades	Duct insulation
Roof, wall, and foundation insulation	Retro-commissioning
High-efficiency windows	**Industrial Sector Measures**
Faucet aerators and low-flow showerheads	Efficient process heating
Pipe insulation	High-efficiency motors and drives
Programmable thermostats	High-efficiency heating, ventilation and air conditioning (HVAC)
In-home energy displays	Efficient lighting

Primarily, they tend to differentiate between technical potential (which looks at technical feasibility without regard to cost or implementation issues), economic potential (which limits the items to those that are cost-effective), and achievable potential (which takes into account economics and the realities of program implementation). This filtering process reduces the amount of energy savings to that which is practical. Table 10-3 highlights some of the important variables used in these studies.

One of the earliest, thorough discussions of the potential for energy efficiency was published in 1990 by Scientific American (Fickett, et al., 1990). As shown in Figure 8-1, the article showed a range of estimates generated by two studies. One by EPRI (EPRI CU-6744) and the second by the Rocky Mountain Institute. While they depicted a range from nearly 25% to over 40% in technical potential, they did highlight many of the important technology areas where the potential lies.

One perspective on the potential savings from implementing end-use initiatives is provided in recent work sponsored by EPRI through Global Energy Partners (GEP). Figure 10-2 is a supply curve showing

Table 11-3. Range of Study Assumptions

Baseline Forecast	Business as Usual (1% per year improvement)	➡	Moderate Increase in Energy Efficiency	➡	Aggressive Energy Efficiency
Appliance Codes & Building Standards	Codes Known & In Place	➡	Increase in Codes Based Upon Known Proposals	➡	Aggressive Assumptions About New Codes
Utility Programs & Initiatives	Continuing	➡	Assuming Modest Growth	➡	Assume All Utilities Aggressively Pursue
Adoption of Technologies	Realistic Potential	➡	Economic Potential	➡	Technical Potential
Efficiency of End-Use Technology	Current Trend Continues	➡	Trend Accelerates	➡	Several Breakthroughs Identified
Price Elasticity of Demand	Behavior Remains Unchanged	➡	Consumers Become Modestly More Responsive	➡	Consumers Become Very Sensitive to Price
Consumers Control Usage	Behavior Remains Unchanged	➡	Consumers Increase Control	➡	Consumers Uniformly Control Use

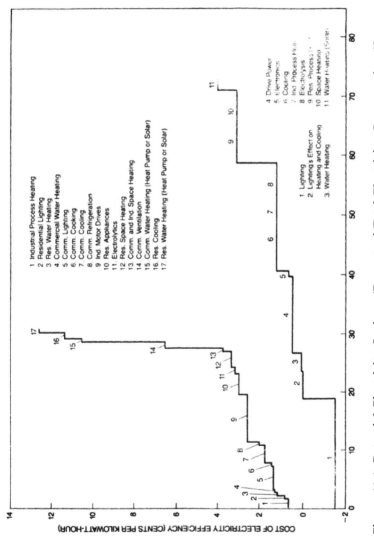

Figure 11-1. Potential Electricity Savings (Percent of Total Electricity Consumption (Source: Fickett, 1990)

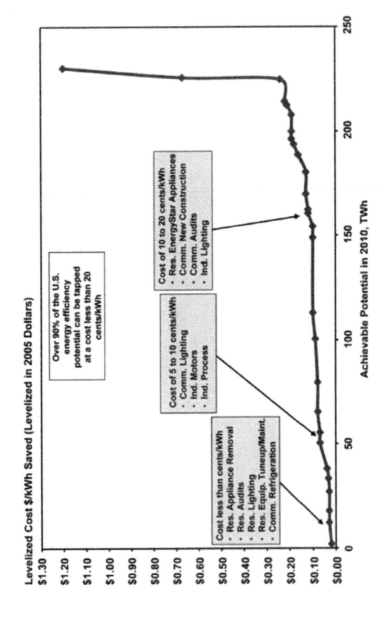

Figure 11-2. Achievable Annual Energy Savings Through Efficiency Improvements, 2010 (Source: Gellings)

the cost of electricity savings in TWh. Note that it is possible to achieve savings of nearly 50 TWh for a cost of less than \$0.05/kWh. Similarly, savings of ~100 TWh are achievable for a cost of \$0.10/kWh to \$0.20/kWh. In fact, savings of nearly 230 TWh (or about 5% of annual U.S. electricity consumption) can be achieved for less than about \$0.20/kWh (Gellings, 2006).

Moreover, it is possible to convert the cost data into a value or benefit. Projecting data from the U.S. DOE/EIA, it's possible to estimate annual electricity consumption in the U.S. at about 4,300 billion kWh. At a retail price of \$0.067 per kWh, the corresponding total revenue can be estimated at about \$290 billion per year. The savings would be about 5% of this value, or ~\$14.5 billion.

Disaggregating this composite supply curve into the three main sectors—residential, industrial, and commercial—reveals some important similarities. For example, the cost of implementing the technologies is initially low but increases steadily, mirrored by the continuous increase in the potential savings in energy and power. All three sectors exhibit this pattern, and all eventually reach a point at which the marginal cost of saving the next kilowatt-hour or the next kilowatt increases dramatically. (Figure 11-3 shows energy savings for all three sectors.) Once this "knee" in the curve is reached, energy efficiency programs will no longer be cost-effective.

But the low-cost portions of the supply curves offer many potential low-cost options for improving efficiency and demand response. Typical efficiency improvements include:

- Removal of outdated appliances
- Weatherization of the building shell
- Advanced refrigeration in commercial buildings
- Residential lighting improvements
- HVAC tune-ups and maintenance

All these technologies can be deployed for a cost of less then \$0.05 per kWh. For a cost of \$0.05/kWh to \$0.10/kWh, additional energy savings include:

- Commercial lighting improvements
- Industrial motors and drives
- Industrial electrotechnologies.

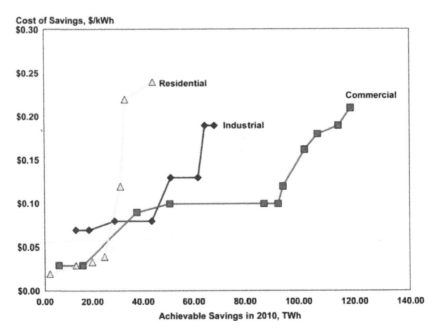

Figure 11-3. Energy Efficiency Savings for Residential, Commercial, and Industrial Sectors (Source: EPRI Summer Seminar, 2006)

In its recent study on energy efficiency, EPRI (Report 1016987) reviewed several other studies on the potential for energy efficiency. A summary of six of these are highlighted here.

The American Council of Energy Efficient Economy conducted a study to estimate the economic and achievable potential for a number of northeastern states over the period 2005 to 2020 (ACEEE, 2006). The economic potential estimates range between 26 to 31% for the time period covered. The achievable potential was estimated to be two-thirds of the economic potential.

- The California Energy Commission (CEC) prepared a staff report which provided estimates of the technical, economic, and achievable potentials for the state of California in the year 2016 (CEC, 2007). These savings estimates were aggregated from individual utility data from all utilities in the state. Results indicated that the technical potential is 23%, economic potential is at 18%, and the achievable potential is at 9%.

- The Midwest Energy Efficiency Alliance (MEEA) estimated both technical and achievable potential for the residential sector only in the Midwest region (MEEA, 2006). The potential estimates were provided for the year 2025. The technical potential for the residential sector is estimated at approximately 24%, while the achievable potential estimate is close to 10%.

- The Western Governor's Association estimated the energy savings potential for 18 western states that below to the WGA (WGA, 2006). The study estimated achievable potential for the three years 2010, 2015 and 2020 at 7%, 14%, and 20%, respectively.

- ACEEE conducted a study for the state of Florida estimating the achievable potential in the state for the years 2013 and 2023. They estimated the achievable potential to be 6.6% for 2013 and 20% for 2023.

- McKinsey and Company conducted a study in order to estimate the costs and potentials of different options to reduce or prevent greenhouse gas emissions within the U.S. over a 25-year period (McKinsey, 2007). The team concluded that energy-efficiency programs and policies directed at factories, commercial buildings, and homes could contribute up to 15% of reduced carbon emissions by 2030.

HISTORIC PERSPECTIVE ON ENERGY EFFICIENCY

California has demonstrated the viability of energy-efficiency measures over many decades. Figure 11-4 depicts per capita electricity sales (kWh/person) comparing California with the United States.

The California Energy Commission (CEC) has estimated that the savings due to the application of energy-efficiency programs and standards in California between 1976 and 2003 are roughly 15% of annual electricity consumption (see Figure 11-5).

CEC estimated the average cost of efficiency measures for the period 2000 to 2004 to be approximately of $0.03/kWh saved. Similarly, the National Action Plan for Energy Efficiency cites a common rule of thumb that "many energy-efficiency programs have an average life cycle of $0.03/kWh saved, which is 50 to 75% of the typical cost of new power sources."

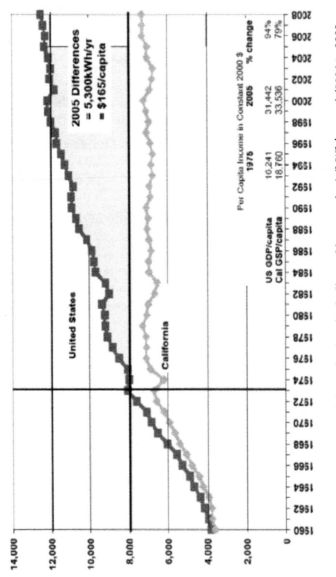

Figure 11-4. Per Capital Electricity Sales (not including self-generation) (kWh/person) (2006 to 2008 are forecast data) (Source: Rosenfield, 2008)

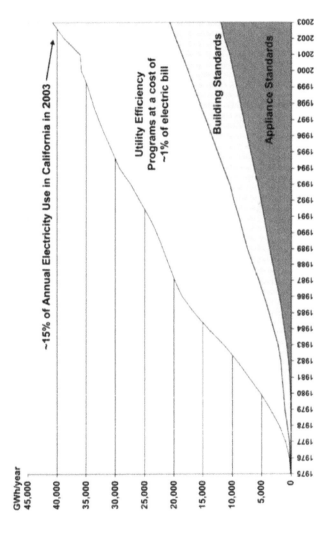

Figure 11-5. Annual Energy Savings from California Efficiency Programs and Standards (Source: Rosenfeld, 2005)

Other studies by the CEC show the dramatic effect of appliance standards on the efficiency of refrigerators in the U.S. over the 30-year span from 1972 to 2002 (see Figure 11-6). While the size of refrigerators has continued to grow, both energy use and price per unit have declined by nearly two-thirds.

One perspective from Europe, show, in Figure 11-7, comes to a similar conclusion of large-scale savings over the last 35 years. The figure indicates that the European Union considers "negajoules" as a viable element in the portfolio of strategies to response to anticipated demand growth. By extrapolation, the "negajoules" shown in the graph, representing avoided consumption, are estimated to reach on the order of 13% of total consumption by 2020 if aggressive efficiency measures are implemented.

References

"Assessment of Achievable Potential from Energy Efficiency and Demand Response Programs in the U.S.: (2010-2030)," EPRI, Palo Alto, CA: 2009. 1016987.

"Energy Efficiency's Role in a Carbon Cap-and-Trade System: Modeling Results from the Regional Greenhouse Gas Initiative," American Council of Energy Efficient Economy (ACEEE), Report No. E064, May 2006.

"Statewide Energy Efficiency Potential Estimates and Targets for California Utilities," California Energy Commission (CEC), Draft Staff Report, CEC-200-2007-019-SD, August 2007.

"Midwest Residential Market Assessment and DSM Potential Study," commissioned by the Midwest Energy Efficiency Alliance (MEEA), March 2006.

"Energy Efficiency Task Force Report by the Western Governor's Association (WGA)—Clean and Diversified Energy Initiation," January 2006.

"Potential for Energy Efficiency and Renewable Energy to Meet Florida's Growing Energy Demands," ACEEE Report No. E072, June 2007.

"Reducing U.S. Greenhouse Gas Emission: How Much at What Cost?" McKinsey & Company, U.S. Greenhouse Gas Abatement Mapping Initiative Executive Report, December 2007.

"Assessment of U.S. Electric End-Use Energy Efficiency Potential, Clark W. Gellings (EPRI), Greg Wikler and Debyani Ghose (Global Energy Partners, LLC), Lafayette, CA, *Electricity Journal*, 2006

"Advancing the Efficiency of Electric Utilization: Prices to Devices[SM]," Background Paper, EPRI Summer Seminar, 2006.

"Efficient Use of Electricity," A.P. Fickett, C.W. Gellings, and A.B. Lovins, *Scientific America*, September 1990.

"Efficient Electricity Use: Estimates of Maximum Energy Savings, EPRI, Palo Alto, CA. March 1990, CU-6746.

"Past and Current Efficiency Successes and Future Plans," ACEEE: Energy Efficiency as a Resource, Berkeley, CA, September 26, 2005.

"The 2009 State Energy Efficiency Scorecard," American Council for an Energy-Efficient Economy," October 2009.

"Energy Efficiency in California," Arthur Rosenfeld, Climate Group Breakfast Preceding Gov. Schwarzenegger's Climate Summit, November 18, 2008.

"U.S. Utilities Spent $5.3 Billion on Energy Efficiency Programs in 2008," U.S. Department of Energy, *EERE News*, February 24, 2010.

"Demand Response and Energy Efficiency: The Long View," Ahmad Faruqui, Presentation at Goldman Sachs 10th Annual Power and Utility Conference, New York, NY, August 12, 2010.

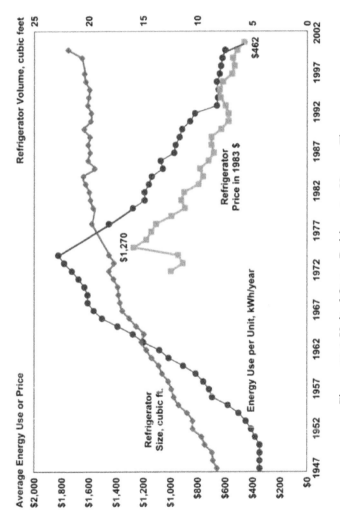

Figure 11-6. United States Refrigerator Use vs. Time

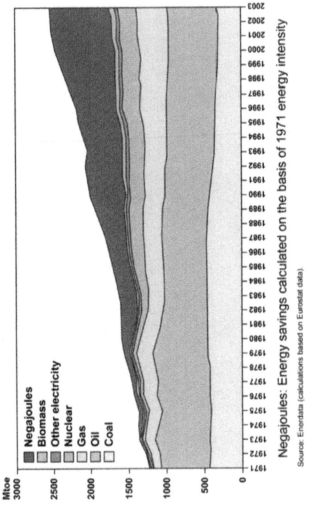

Figure 11-7. Development of Primary Energy Demand and Avoided Energy Use (Negajoules) in the EU 24

Chapter 12

Demand Response

Rationalizing the pattern and amount of electricity use to the wholesale electricity market would reduce electricity prices and increase available capacity. In particular, end-use energy efficiency and the coordination of dynamic changes in energy usage via demand response (DR) remains a critically underutilized resource in the United States. "Demand response" is represented by a shifting of the pattern of load. DR has a small impact on cumulative energy reduction but a large role in enhancing system economics and reliability. This resource will become strategically more important as carbon constraints and affordability of energy create greater economic challenges to energy companies and consumers. Although the potential size of demand-side resources is a matter of ongoing debate and analysis, there are estimates in the range of 10 to 25% of total U.S. electricity consumption (for energy efficiency) and an additional 5% or more of peak demand. The upward bound on this potential is likely to grow as technology advances and as regulators and policy makers elevate its strategic priority.

Results from the 2008 Federal Energy Regulatory Commission (FERC) survey indicate that about 8% of customers in the U.S. are on some form of demand response program, either incentive-based demand response or a time-based rate. There have been changes in customer participation in these programs. For example, the number of customers on real-time pricing and critical peak pricing programs increased since 2006. Regarding the demand response resource contribution nationally, FERC staff estimated a potential peak load reduction of about 14 GW. This represents approximately 5.8% of the total U.S. projected electricity demand for summer 2008, and a 9% increase in potential from the 38 GW estimate of potential peak load reduction in 2006.

The FERC survey focused on incentive-based and time-based demand response programs. Incentive-based programs involve an inducement or incentive for customers to modify their electricity consumption.

This is in contrast to time-based rates. Incentive-based demand response program provide a direct means for controlling load and are, therefore, used by load-serving entities, electric utilities, or grid operators to manage costs and maintain reliability, especially in emergency conditions when immediate and predictable demand response is required (FERC, 2003).

According to the U.S. DOE (DOE, 2006), the kinds of incentive-based programs include:

- **Direct load control**: A demand response activity in which the program sponsor remotely shuts down or cycles a customer's electrical equipment.

- **Interruptible/curtailable rates**: Curtailment options integrated into retail rates that provide rate discount or bill credit for agreeing to reduce load.

- **Emergency demand response**: Emergency demand response programs provide incentive payments to customers for voluntarily reducing their loads during reliability-triggered events.

- **Capacity market programs**: In capacity market programs, customers commit to providing pre-specified load reductions when system contingencies arise.

- **Demand bidding/buyback programs**: Demand bidding/buyback programs encourage large customers to provide load reductions at a price at which they are willing to be curtailed.

- **Ancillary services market programs**: Demand response programs in which customers bid load reductions in ancillary services markets.

Direct load control programs are the most common type of demand response programs. They are typically operated to balance supply and demand system peak. There have been direct load control programs with meter-based and consumer-based equipment controls since the 1960s. FERC survey responses indicate an approximate 3.5% growth in customer enrollment in direct load control programs since 2006, up from 4.95 million customers to 5.13 million.

In addition to incentive-based demand response programs, demand response is also accomplished through the use of direct price signals associated with time-based rates.

In the 2008 FERC survey, the number of entities that reported offering time-based rates totaled 503 (see Table 9-1).

Table 12-1. Number of Entities Offering Time-Based Rates (Source: FERC, 2008)

Time-Based Rate	Number of Entities (2006 Survey)	Number of Entities (2008 Survey)
Time-of-Use Rates	366	315
Real-Time Pricing	60	100
Critical-Peak Pricing	36	88
Total	462	503

Demand response is enabled by dynamic systems. "Dynamic systems" represent the future of networked, smart, end-use devices interacting with the marketplace for electricity and other consumer-based services. Market interaction includes either sending direct "prices to devicesSM" or making price signals available to relevant information technology and consumer electronic devices. This area may have substantial impacts on system reliability, customer value, energy savings, and CO_2 reductions (EPRI 2006 Summer Seminar).

While there is value in demand response programs, it is not clear how that value of DR-ready appliances, enabled either directly or through information technology and consumer electronics, translates to consumers (i.e., whether they ultimately lead to energy or cost savings for the consumer). If, in the end, that value proposition does not exist or does not exist in a consistent manner across a range of products, then encouraging consumers to purchase products with a DR-ready label will not be successful.

Promoting demand response differs from promoting energy efficiency in that instead of saving cost by saving energy, the principle value in demand response is that it mitigates operational risks (e.g., price risk and reliability risk faced in day-to-day system operations), These create risks of high wholesale costs at uncertain times throughout the year (e.g., the California Energy Crisis occurred in the winter non-peak demand season). Thus demand response addresses risks of high whole-sale costs for electricity, while energy efficiency investments by utili-ties can be designed to maximize overall capacity savings, they are not designed to provide a dynamic response to address operational risks. Mitigating risks of uncertain wholesale costs, risks faced in day-to-day system operations, using resource adequacy and other planning type measures can be more expensive than solutions that equip the demand-side to mitigate the risk with demand response. The costs for addressing risks are ultimately passed down to retail customers, even if retail rates are flat for the time being. So energy efficiency (which delivers the same or improved level of service to customers with fewer kWh energy sav-ings) does not necessarily achieve as much overall cost savings to the electricity system as DR in market environments exposed to such risks. Consequently, the usefulness of the DR-ready concept depends on the value of mitigating reliability and price risks, which may differ across regional environments.

DEFINITIONS

The word *appliance* has different meanings to product manufactur-ers and laymen. U.S. DOE and EPRI use appliance broadly to include white goods, small appliances, space conditioners, electric water heat-ers, thermostats, battery chargers, and perhaps other equipment.

The terms *information technology* and *consumer electronics* used in this chapter refer to a variety of digital devices which can be used to en-able end-use energy-consuming devices and appliances to be DR-ready. These include advanced metering infrastructures, controllable thermo-stats, energy monitors and the like.

The term *demand response* has often been applied to refer to the managing of peak load. However, future demand response-ready appli-ances could more broadly enable many additional services to support the electric power grid through market opportunities, including ancil-lary services.

Demand response-ready appliances and devices refers to appliances, in-formation technology and consumer electronics that are readily applied to demand-response programs after those devices become purchased by consumers.

The demand response-ready appliance interface is the junction between the appliance and a higher level system that communicates with the appliance. This higher-level system could well be devices involving information technology and consumer electronics. This interface is where the demand response request is received by the appliance and acknowledgements and other information may be transmitted. (For example, the interface may become reduced to the point where it can be communicated by the states of a small number of pins. Under this example, the process initiated by this white paper could standardize the interface hardware and the accepted interpretations for these pins.)

Demand response aggregators are firms or institutions which engage groups of customers to aggregate their participation in demand response programs.

BACKGROUND

A widely held belief among utility planners, policymakers, regulators and observers is that, if consumers' cost of electricity reflected the impact their demand (time, pattern and amount) actually had on the power system, they would optimize that demand so as to minimize the cost to themselves while coincidentally minimizing societal costs. Demand response (DR) programs and associated technologies are intended to enable consumers to participate in such programs and activities—either voluntarily or through mandatory arrangements. DR programs can be based on a variety of possible incentives or technology arrangements wherein consumers are directed to or incented to modify their demand at various times.

DR programs and technologies typically include those involving either passive price incentives; dynamic price incentives; voluntary load control; or involuntary load control.

Passive price incentives are those offered by time-of-day, time-of-use and off-peak rates which offer the consumers some price differential through time periods. Typically, these time periods are fixed and vary only by time of day and possibly by weekends vs. weekdays. In theory, reinforced by communication, consumers modify their demand for electricity in various time periods in response to their understanding of the potential impact on their electric bill. These systems and the incentives involved are passive and produce the least modification in consumer behavior in coordination with grid needs.

Dynamic price incentives are those which are implemented through the use of technology which allows communication directly with consumers and/or their building systems and appliances. Technologies enabling dynamic pricing or dynamic systems can take a variety of forms including systems which dispatch day-ahead prices to an "advanced metering infrastructure" which programs the next day's demand for electricity use by major appliances—possibly subject to consuming parameters which are preset. Another dynamic option involves "critical peak pricing" (CPP) wherein the consumer receives a strong price signal which (1) inspires response to reduce demand, (2) directly controls an appliance or device, or (3) penalizes the consumer with a higher bill. Finally, these systems can simply involve a signal which is sent to a monitor, a thermostat, a special alarm or the TV to alert consumers—giving them the option to modify behavior. These price signals can be communicated by utilities or others directly to appliances or through information technology and consumer electronics.

Voluntary load control involves engaging consumers, with or without incentives, to modify their demand in response to some communication such as notification messages. The communication can be via email, phone, fax or special signal. The consumer response is entirely voluntary and may not involve incentives. Manual actuation of the actual energy system is not implied.

Involuntary load control involves the use of hard-wired communications systems wherein the consumer's information technology interface, consumer electronics, or appliances and end-use devices, circuits or substations are directly interrupted. In most cases, these arrangements are part of a special service offering or tariff wherein the consumer receives a financial benefit to allowing control. This type of demand response is among the most valuable, since the utility and system operator are assured of the availability of load reduction. Involuntary is distinguished from voluntary primarily by automation of response through directly controlled use.

In general, energy efficiency is a by-product of all forms of demand response. Numerous studies have shown that when demand response programs are in effect, consumers become more aware of their energy use and tend to be more selective in the purchase of appliances and end-use, energy-consuming devices and in their utilization.

FOUR BUILDING BLOCKS OF DEMAND RESPONSE

An integrated set of four building blocks—communications infrastructure, innovative rates and regulation, innovative markets, and smart "DR-ready" end-use devices—constitutes an emerging demand response infrastructure that will make the dynamic dimension of energy efficiency more robust over time, substantially expanding the potential for energy efficiency and demand response in the broadest sense (Figure 12-1).

The integration of the four building blocks is necessary to realize the full potential of the demand response infrastructure. In the future, effective communication of increasingly larger amounts of data will be required, and market rules and regulatory conditions must exist to incent and enable improved efficiency, demand response, and dynamic systems. Finally, a set of advanced end-use IT or CE technologies with embedded intelligence so as to be "DR-ready" will be required to implement the energy management necessary to achieve dynamic system.

WHERE IS THE BIGGEST IMPACT?

The biggest impact on overall costs for consumers and society are from those where the response of appliances is dynamic through either direct prices-to-devices or indirectly through IT and CE, and those where load control is involuntary. Both of these schemes require

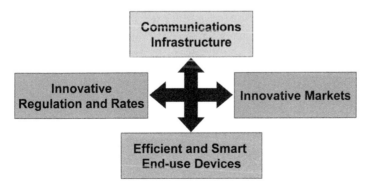

Figure 12-1. The Four Building Blocks of a Demand Response Infrastructure

a communications system either overlaid on the electric power system or available, independently to provide price signals and other information.

Utilities, FERC, DOE, state regulators and others are presently engaged in aggressively pursuing the evaluation and implementation of demand response. While estimates vary, the consensus among experts is that the impact of demand response can reduce the need for peak generation capacity by 5% or more, while reducing consumer bills by as much as 20%, and effecting an overall reduction in electric energy consumption by 4% or more.

Unfortunately, there are three obstacles to enabling ubiquitous demand response in the U.S. These are (1) the ability to communicate, hampered by the multitude of communications protocols used in the utility industry (~152) and the protocols used in the building industry (>28); (2) the degree to which end-use energy-consuming devices, appliances and building energy management systems are "DR-ready"; and/or the application which enables DR-ready information technology and consumer electronics devices. DR-ready is herein defined by the authors as being capable to receive signals (and instructions) and to respond automatically to those signals.

THE POTENTIAL FOR SUMMER PEAK DEMAND SAVINGS FROM UTILITY PROGRAMS

EPRI estimated two types of summer peak demand savings (EPRI 1016987). These included the reduction from energy-efficiency measures which inherently reduce summer peak demand insofar as their usage is coincident to the overall summer peak. In addition, utility demand response programs specifically targeted at peak demand reduction result in additional savings. EPRI estimated that energy efficiency and demand response contribute to an achievable peak demand reduction potential of 157 to 218 GW in 2030, or 14 to 20% of projected U.S. summer peak demand in 2030.

Table 12-2 and Figure 12-2 present the potential peak demand savings.

Demand response programs considered in the EPRI analysis include the following:

Table 12-2. Potential for U.S. Summer Peak Demand Savings (GW) (Source: EPRI 1016987)

Realistic Achievable Potential	2010	2020	2030
Energy Efficiency	1.6	34.8	78.5
Demand Response	16.6	44.4	78.4
Total	18.2	79.2	156.9
Maximum Achievable Potential	**2010**	**2020**	**2030**
Energy Efficiency	10.8	81.7	117.0
Demand Response	29.8	65.9	101.1
Total	40.6	147.6	218.1

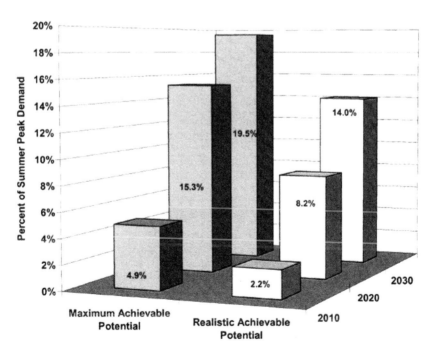

Figure 12-2. Potential for Summer Peak Demand Savings from Energy Efficiency and Demand Response (Source: EPRI 1016987)

- Residential sector: direct load control (DLC) for air conditioning and water heating, dynamic pricing programs including time of use (TOU), critical-peak pricing (CPP), real-time pricing (RTP), and peak-time rebates.

- Commercial sector: direct control load management for air conditioning, lighting, and other uses; interruptible demand (e.g., interruptible, demand bidding, emergency, ancillary services); and dynamic pricing programs.

- Industrial sector: direct control load management for process; interruptible demand (e.g., interruptible, demand bidding, emergency, ancillary services); and dynamic pricing programs (TOU, CPP, RTP).

Based on the EPRI analysis, the range of achievable potential for demand response programs in 2030 is 7 to 9% of peak demand. The expected savings from demand response measures are roughly equal across the three sectors. The three categories of measures—direct load control, dynamic pricing, and interruptible demand—each deliver roughly the same level of savings. Tables 12-3 and 12-4 present the contributions of major types of demand response programs to peak demand reduction for realistic and maximum achievable potentials, respectively.

Figure 12-3 illustrates the realistic achievable potential of demand response for peak demand reduction by sector and program type.

EXAMPLES OF REAL APPLICATIONS

Marriott Marquis Hotel

One example of how effective demand response can be was demonstrated in a project sponsored by the New York State Energy Research and Development Authority (NYSERDA), Consolidated Edison (ConEd), the Empire State Electric Energy Research Corporation (ESEERCO), Pacific Gas and Electric (PG&E), and the Electric Power Research Institute (EPRI). The objective of the study was to develop and demonstrate automated control strategies for commercial building mechanical and lighting systems in response to real-time pricing (RTP) of electricity (EPRI 111365).

The premise of the study was that the potential opportunity for

Table 12-3. Summer Peak Demand Savings from Demand Response

Residential DR	2010	2020	2030
DLC – Central Air Conditioning	3,128	8,194	11,742
DLC – Water Heating	1,431	2,868	3,931
Price Response	1,539	6.,918	10,967
Commercial DR	**2010**	**2020**	**2030**
DLC – Cooling	1,336	3,833	4,822
DLC – Lighting	364	1,049	1,358
DLC – Other	256	824	1,159
Interruptible Demand	4,337	8,806	19,450
Price Response	771	4,018	8,368
Industrial DR	**2010**	**2020**	**2030**
DLC – Process	413	1,124	2,245
Interruptible Demand	2,550	3,973	8,701
Price Response	515	2,765	5,697
TOTAL	**16,639**	**44,372**	**78,441**
Percentage of Peak	**2.0%**	**4.6%**	**7.0%**

Table 12-4. Summer Peak Demand Savings from Demand Response Maximum Achievable Potential (MW) (Source: EPRI 1016987)

Residential DR	2010	2020	2030
DLC – Central Air Conditioning	4,119	9,498	12,558
DLC – Water Heating	1,960	3,473	4,503
Price Response	4,318	13,122	16,093
Commercial DR	**2010**	**2020**	**2030**
DLC – Cooling	1,766	4,309	5,099
DLC – Lighting	516	1,377	1,698
DLC – Other	508	1,316	1,623
Interruptible Demand	8,536	13,680	26,410
Price Response	2,180	7,600	12,418
Industrial DR	**2010**	**2020**	**2030**
DLC – Process	824	1,826	3,129
Interruptible Demand	3,572	4,554	9,142
Price Response	1,451	5,154	8,422
TOTAL	**29,750**	**65,910**	**101,093**
Percentage of Peak	**3.6%**	**6.8%**	**9.1%**

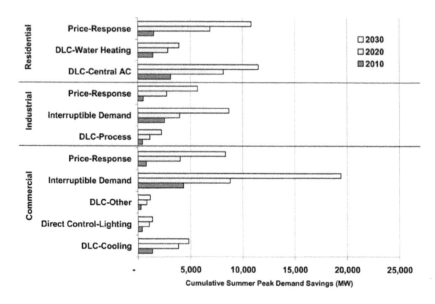

Figure 12-3. Realistic Achievable Potential for U.S. from Demand Response

electric load shedding/shifting of building systems in response to real-time pricing can be significant, providing a cost-reduction benefit for commercial building customers as well as a load-reduction benefit for their electric utilities. In this project, EPRI developed automated energy control strategies for application to typical commercial building HVAC systems (i.e., air handling system fans, pumps, etc.) as well as for other major electrical loads (i.e., lighting, large appliances, etc). The control strategies were configured to respond to real-time prices while also taking into account other user-defined criteria (i.e., occupancy schedules, space temperatures, and other building data). Automated RTP load control technology can enable shedding and/or shifting of the electric load from building systems (typical durations might be from one to three hours) during high-pricing periods. This system enabled the commercial building sector to take better advantage of RTP as an important demand-side management option while helping reduce a utility's reliance on high-cost or purchased energy during peak demand periods.

These automated RTP control strategies were installed at the Marriott Marquis Hotel in New York City, and field performance data were collected from mid-1993 through mid-1995. The results of this program indicate that automated RTP control strategies are feasible for achieving appreciable cost savings and load reductions without adversely affecting worker and occupant comfort. Results were as follow:

- Up to 150 kW of electrical load was curtailed during high-price periods.

- By responding automatically to RTP prices, the building owner saved more than $300,000 in two years of operation.

- Occupant comfort or service in the building was never compromised.

The constituents of the typical monthly electrical cost savings for the Marriott site are shown in Table 12-5.

The results indicate that automated control strategies can provide significant energy and cost savings for a building owner and can also provide significant benefits to the electric utility in the form of reduced electric demand during high-price periods.

World Financial Center

Another similar project was also sponsored by the New York State Energy Research and Development Authority (NYSERDA) and the Electric Power Research Institute (EPRI). The objective of the project was to develop and demonstrate automated control strategies for commercial building ventilation systems in response to real-time pricing (RTP) of electricity.

Table 12-5. Typical Cost Savings Results (Source: EPRI 111365)

RTP-Controlled Load Category	Contribution to Monthly Savings (% of total savings)
Air handling units	60% to 70%, or more
Load reduction at chillers	Up to 30% (in cooling season)
Exhaust fans	10% to 20%
Lighting controls	Up to 2%
Miscellaneous loads	Up to 2%

Obviously, the potential for electric load shedding/shifting of building ventilation systems in response to real-time pricing can be significant. These actions can provide a cost-reduction opportunity for commercial building customers as well as a load-reduction opportunity for their electric utilities. This project demonstrated an integrated sensor/control system for RTP that has the capability to make building ventilation systems more energy-efficient while maintaining prescribed levels of indoor air quality.

Under this project, Honeywell implemented new strategies to provide demand-controlled operation of outdoor air ventilation equipment. These strategies were designed for application to typical commercial building air handling systems in response to real-time pricing of electricity. Honeywell integrated RTP control strategies with a prototype sensor system that measured indoor air carbon dioxide (CO_2) and volatile organic compound (VOC) levels to ensure that prescribed levels of indoor air quality were maintained in the occupied spaces. Honeywell added this demand-controlled ventilation capability to the existing automated RTP load control technology to enable shedding and/or shifting of the electric load from ventilation equipment. This RTP control strategy involved reducing ventilation levels (typical durations might be from one to three hours) during high-RTP periods.

These ventilation control strategies were installed at the World Financial Center (WFC) in New York City, and field performance data were collected during the last quarter of 1996. The results of this program indicated that the RTP control strategy is feasible for achieving cost savings while maintaining prescribed levels of indoor air quality (determined through measured CO_2 levels).

In addition, performance data were collected on automated control of the conventional building loads (fan motors and lighting). The measured data were analyzed, and energy and cost savings were assessed. Highlights of the measured results for the field monitoring period (July 1995 through November 1996) at the WFC included a reduction of up to 450 kW of demand during peak periods without compromising occupant comfort or service.

The constituents of the typical monthly electric cost savings for the Winter Garden area of the WFC are shown in Table 12-6.

Performance results indicate that the RTP control strategy is feasible for achieving cost savings while maintaining prescribed levels of indoor air quality (determined through measured CO_2 levels).

Table 12-6. Typical Cost Savings (Source: EPRI 109117)

RTP-Controlled Load Category	Contribution to Monthly Savings (% of total savings)
Air handling units AC-1W and AC-8W	Up to 5%
Load reduction at chillers	Up to 50% (in cooling season)
Other primary air handling units	Up to 25%
Lighting controls	Up to 25%

HOW TO ENABLE DEMAND RESPONSE

In order to enable demand response, one or more of the following are required: A communications infrastructure; innovative markets; innovative regulation and rates; and smart "DR-ready" end-use devices. The focus of this paper is to discuss methods of fostering an environment that enables end-use devices to come to consumers DR-ready either directly or through enabling information technology of consumer electronic devices.

There are potentially two categories of what we could call end-use devices which define two separate approaches. These are (1) to focus on appliances and related energy-consuming end-use devices; and/or (2) to focus on electronic devices including information technology and consumer electronics (IT and CE). There are differences in orientation which go with each of these approaches (Nordman, 2008).

DR-ready devices could allow direct interface with the power system. In today's world, the DR programs that exist are implemented by going door to door and retrofitting end-use equipment or installing custom-made equipment. If an existing program is to be expanded, then more costly visits need to be made. To make demand response ubiquitous, the interface between the power system and end-use devices needs to be standardized and then embedded in the end-use appliances and devices.

Several industry stakeholders recognize the value in standardizing functionality and communication protocols to assist in the development of DR-ready products and are interested in continuing to discuss how best to approach this area. The four organizations anticipate playing a role in helping to facilitate agreement around greater standardization.

Beyond standardization of important protocols, it is unclear as to the best program approach to enable ubiquitous demand response. This could be focused at appliances and energy-consuming end-use devices and/or information technology and consumer electronic devices. Several options can be considered including a national consumer labeling program for DR-ready devices, a collaborative incentive approach, a program offering incentives upstream from the consumer, or some combination thereof. The purpose of this paper is to stimulate discussion and interest in evaluating a broad range of program approaches, and ultimately, to outline a scope, plan and budget to pursue the most advantageous ones.

In the future, price and other electricity-related information may well come via the electricity distribution system, but it may also come through other data and network methods. Both channels offer possibilities. It may be premature at this point to pick one or the other as the "winner" in advance. Prices could come through the grid or through other communications media. Local building networks will always provide the "last hop" even when price signal comes through the price signal comes through the electricity system, so that from the appliance perspective, the original source of the price may not be of importance. Thus, there may be no direct interface between the power system and the end-use device. A standard interface (or a few) does not need to be defined, but that definition may not be tied to the grid (Eustis, et al., 2007).

In summary, Figure 12-4 illustrates the key approaches to making consumer buildings, facilities and energy-consuming systems DR-ready. DR-ready can be directed to appliances (actually both energy-consuming devices and appliances), to the information technologies and consumer electronics (IT & CE) which interface with appliances or, as today, with the installation of utility-control devices (relays, load management hardware and advanced meters).

There are many mechanisms to enable DR-ready appliances and IT and CE devices. Broadly, they include the ENERGY STAR approach, a mechanism called Energy Smart and other market mechanisms described herein.

WHAT DOES DR-READY ENTAIL?

In order for various end-use products to be DR-ready, the industry would need to agree to common communications protocols or interfac-

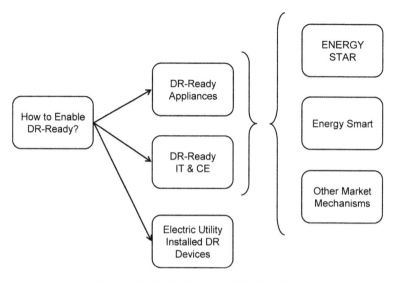

Figure 12-4. The Path to "DR-Ready"

es to enable ubiquitous communication between some communications paths involving wireless or wired communication, or the advanced metering infrastructure (AMI), as well as building or consumer portals, and the appliances, devices or building energy management systems (EMS). Enabling this protocol or interface would entail either embedding a "chip" in all major end-use electricity consuming appliances and devices so as to allow them to receive and respond to signals, or enabling information technology and consumer electronics to receive those price signals and appropriately control appliances and devices. Either of such capabilities would make the end uses "DR-ready."

WHY AREN'T TODAY'S APPLIANCES, INFORMATION TECHNOLOGY AND CONSUMER ELECTRONICS DR-READY?

While there is widespread interest in enabling DR, there are numerous obstacles to a situation where distribution utilities install advanced infrastructures and consumers enjoy their benefits. The issue simply stated is "who pays and who gains." For example, some utilities are encouraged and incented to adapt AMI—most are not; some consumers are offered incentives, rewards and lower bills to participate in DR programs—most are not; and nowhere are there obvious incentives for the manufacturers of end-use energy consuming devices and appliances, information technology to offer DR-ready devices.

The key to enabling ubiquitous DR is in encouraging or incenting the manufacturers of end-use devices and appliances to make their products DR-ready.

ALTERNATIVES TO ENABLING DR-READY END-USE PRODUCTS

DR-ready end-use products will ultimately only come to market if consumers can identify and realize value through their adoption. Few consumers "see" electricity. They don't know where it comes from; they don't know how it is converted into the services they desire (warmth, cooling, lighting, refrigeration, cooked food, etc.). Some consumer segments are beginning to recognize the sensitivity of their usage on the environment and on their bills. So energy efficiency is increasingly a factor in consumer purchase decisions.

Encouraging adoption of DR-ready end-use products differs from promoting energy-efficient appliances and devices in that instead of the consumer saving cost by saving energy, demand response mitigates risks of high energy costs at uncertain times during the year. Mitigating risks of uncertain wholesale costs, risks faced in day-to-day system operations, using resource adequacy and other planning type measures can be much more expensive than solutions that equip the demand-side to mitigate the risk with demand response. The costs for addressing risks are ultimately passed down to retail customers, even if retail rates are flat for the time being. So energy efficiency (which achieves kWh savings) does not necessarily achieve as much savings as DR in a market environment exposed to such risks.

Figure 12-5 illustrates the array of possible participants in the value chain for demand response. Demand-response value can originate principally in the wholesale electricity market or within the distribution utility as part of the retail service provider function. The value of demand response is ultimately seen by the consumer in the form of lower average cost where time differentiated pricing is available, incentives for participating in demand response programs and/or enhanced reliability. The consumer realizes this value by their interaction with their service provider or by engaging with a demand-response aggregator.

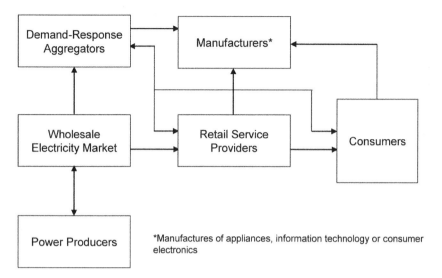

Figure 12-5. Value Chain Participants in Demand Response

The demand-response value in mitigating risk and reducing wholesale electricity cost can be passed on to either the retail service provider, the demand-response aggregator, or in special circumstances where certain ancillary service arrangements encourage the use of demand response to facilitate production operations, to the power producers.

Distribution utilities and retail service providers can use demand response to perform the same function as an aggregator and reduce overall wholesale prices they pay and/or enhance reliability. In addition, the distribution asset owner/operator may use demand response to enhance asset management, and in so doing, reduce maintenance costs and enhance reliability.

For DR-ready appliances to be available in the marketplace, appliance manufacturers need to enable their products. It is likely that making appliances and devices DR-ready may add a slight incremental cost to the product. However, adding these features may well facilitate incorporating other functionality thereby virtually masking any incremental cost. Manufacturers can be stimulated to make products DR-ready either by perception that doing so will increase market share, by appliance standards, or by responding to incentives. These incentives can flow from either demand-response aggregators, retail service providers or consumers through their demand for DR-ready appliances and devices.

Among the barriers to achieving widespread demand response is the issue of designation and certification to encourage manufacturers to develop DR-ready products that interface to utility communications. As illustrated in Figure 12-6, the evolution of the electric "smart grid" infers the integration of utility system communications with major end-uses, enabling a plethora of new resources that could support grid operations. The interface must be seamless and allow interoperability to be a viable solution for mass deployment.

This chapter assumes that a market-based program to stimulate manufacturers is the preferred approach. Therefore, for widespread deployment of end-use, energy-consuming devices and appliances and building energy management systems that are DR-ready, mechanisms must be in place to either (1) stimulate consumer demand for DR-ready products; or (2) facilitate the flow of incentives or purchases from providers or aggregators to manufacturers.

Some industry experts believe that "smart grids" should, for the most part, end at the meter. This does not mean there will never be grid-oriented information that reaches through the meter to individual devices, but that this could be the exception rather than the rule, and should not be a design constraint on the grid, on building networks, or individual devices. DR is a leading method for rationalizing our electricity system, but direct use of prices (to devices) is the more important long-run method, so that the DR strategy should be designed to enable direct price responsiveness as early as possible, perhaps right at the very beginning. As such, enabling responsive information technology or consumer electronics may be the most effective strategy.

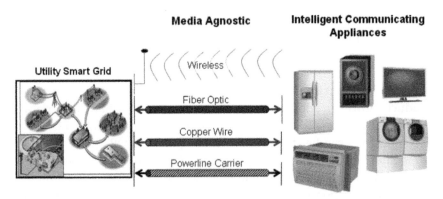

Figure 12-6. The Utility Smart Grid Interfaces with Intelligent Appliances, Consumer Electronics, and Other End Uses

There are numerous potential alternatives mechanisms to use to accomplish this. A few are outlined here to stimulate discussion. These include:

- A collaborative incentive approach (similar to the 1990s "golden carrot" approach).
- A national consumer labeling program for DR-ready devices.
- A program offering incentives upstream from the consumer.

Collaborative Incentive Approach

During the 1990s, EPRI and the Consortium for Energy Efficiency (CEE) worked together with a number of utilities to formulate and launch a "golden carrot" program. That program was targeted at large, highly efficient, residential refrigerators. It became the framework for a number of subsequent efforts. The term is borrowed here for discussion purposes.

This approach would engage retail service providers and possibly aggregators in a consortium which would establish guidelines for DR-ready appliances and devices. Consortium members would guarantee to only allow qualifying appliances and devices to be part of demand-response programs they sponsored. In parallel, they would promote consumer awareness of and interest in these DR-ready products. In addition, they may provide various incentives. Some consortium members could also sell these appliances and devices.

A National Consumer Labeling Program Approach

ENERGYSmart (working title) is a voluntary program engaging manufacturers of devices. The objective of such a program would also be to establish DR-ready certification and designation. This would entail first identifying the stakeholders, then identifying technical requirements and specifications to embed into the certification, and developing a roadmap of viable end-use applications. With voluntary participation from a majority of manufacturers, this should foster mass development of DR-ready equipment and their deployment in buildings.

The challenge to ENERGYSmart would be to decide who the target audience is and then to build brand recognition aimed at that audience.

Recognition and awareness of the ENERGYSmart Program among electricity consumers could be greatly enhanced by utility support through communication, customer contact and incentives. Placing the full weight of the electric energy service provider behind this effort could prove quite effective. In this case, the target audience for the brand would be both the consumer and the service provider or demand-response aggregator.

As the DR-ready concept reaches fruition, the economics of DR programs will improve. For example, utility representatives (e.g., retrofit installers) will not need to visit a home every time a communications device or demand responsive appliance is to be added. Today, separate visits are needed to modify devices to make them DR-compatible: programmable thermostats, water heaters, pool pumps, etc. As a result, the reach of DR programs will increase. This could have the side benefit of increasing consumer satisfaction with their energy purchase experience.

In parallel, the cost-effectiveness of demand-side resources will improve. The amount of controllable load will increase, enhancing the reliability of the power system and the dependability of DR programs themselves. Demand-side resources such as DR will become increasingly competitive, vis-à-vis supply-side options, initially for peak power and broadening into intermediate power as DR-ready appliances and devices begin to include features such as energy storage.

Ultimately, widespread demand-side integration will result in "Ubiquitous Demand Response" which transforms purchase patterns of demand, enables enhanced use of distributed resources, and increases the benefits that energy service providers and their customers receive.

It is envisioned that the ENERGYSmart approach requires an extensive effort over a long period of time. Similar programs have been developed which target energy efficiency. These programs typically experience a modest beginning and build in effectiveness over a period of time. A national labeling program aimed at DR-ready devices could transform the marketplace once the customer value proposition for DR-ready investments becomes clear.

Network requirements play an increasing role in ENERGYSmart specifications for electronic products (computers and imaging equipment probably the best current examples). The way these are developed and treated may be a good model for DR in the future.

It will take several years before we can see any DR capability installed as a standard feature in "appliances," due to the need to define the software interfaces and add the network hardware. However, we can start with electronic devices that already have the network communication necessary, and already have the ability to update their software after purchase. That is, we could potentially bring millions of devices into some DR system in a matter of months rather than years. This could bring valuable practical experience with technology and for the public at an early stage. The amount of power reduction available may well be much less than with more traditional appliances, so your emphasis on those devices is, of course, very well placed. However, the "electronics first" strategy may help get quicker and better success in the other areas.

A key point about electronics is that the network connectivity (hardware, software, and user interface) is all bought and created for other functional purposes of the product, not on the tab of energy or DR specifically. DR does not need to spend money to "add chips" to products—the hardware is already there. This brings the incremental cost to nearly or exactly zero, greatly increasing the east of adding the feature to products at time of manufacture.

A POSSIBLE APPROACH TOWARD IMPLEMENTATION OF DR-READY PROGRAMS

To stimulate discussion regarding the key steps that would need to be taken to include DR requirements in any of the approaches mentioned, six key activities need to be addressed. Some of these can take place in parallel, and others in sequence.

1.   Identify underlying drivers and interests among key stakeholders.
2.   Determine state of industry and technology today.
3.   Define product attributes that warrant "DR-ready" designation.
4.   Build coalition with strategic partnerships.
5.   Develop roadmap of target products.
6.   Develop standards for exchanging information with smart appliances.

IDENTIFY UNDERLYING DRIVERS AND INTERESTS AMONG KEY STAKEHOLDERS

An initial assessment of the potential key stakeholders and their primary interests and drivers has resulted in identification of several groups. They include the following:

- Energy service providers, vertically integrated utilities, distribution utilities, DR aggregators, independent system operators (ISOs) and regional transmission operators and their trade associations will all be interested in the functional requirements and specifications that are compatible with current and future demand response programs. This could involve addressing issues such as:
  — Communications protocols
  — Standards
  — Performance metrics
  — Telemetry requirements

These entities would be engaged both individually and through their trade associations including the Edison Electric Institute (EEI), the National Rural Electric Cooperation Association (NRECA), and the American Public Power Association (APPA).

There is still uncertainty in how these entities will be linked in the DR value chains. For example, aggregators may not play much or at all in the residential sector, but can be essential in the commercial sector.

- Federal and state agencies such as the U.S. Department of Energy, the Environmental Protection Agency, state energy agencies and public utility commissions will be interested in the substantial public benefit which can accrue from demand response.

- Consumers and their representative organizations will benefit from the increased reliability of the power system as well as the enhanced features which the associated energy management systems will likely provide. In addition, a DR-ready system will likely provide additional automation and convenience to the consumer. Finally, the consumer will have the ability to receive better real-time information on energy use through monitors and display devices.

It should be noted that there are two distinct "value propositions" for consumers—one for acquiring the capability in the product, and a second for using it.

- Equipment manufacturers have the opportunity to expand market share and to label their energy-consuming devices and appliances as DR-ready, thereby enhancing their competitive advantage.

The National Electrical Manufacturers Association (NEMA) and the Consumer Electronics Association (CEA) may be particularly helpful in outreach to these companies.

- In the near term, EPA is working with NEMA and their manufacturing partners to develop a new ENERGY STAR programmable thermostat specification that will identify and reinforce energy saving behavior in the consumer. New options that they hope to include in this new specification include managing and coordinating HVAC home controls that affect thermal comfort and air quality and allow consumers to use time-of-use pricing. Part of the current NEMA standards setting process for programmable thermostats is their effort to develop a "strawman" proposal for criteria for a new specification. This proposal is anticipated to be crafted by the fall of 2008.

- The distributors and retailers of devices are anxious to meet consumer demands for products and services. As the DR-ready concept expands, this could enhance their success in the market. This could be enhanced by the energy service providers' involvement. Other appliance equipment organizations may also play a role, including the American Home Appliance Manufacturing Association (AHAM), and the Air Conditioning and Refrigeration Institute (ARI).

- Architects and consulting engineers have an obligation to offer their clients designs and specifications which provide them the greatest value. Therefore, to the extent a DR STAR-type program would be available, they would be anxious to employ it. Associates with potential interests linked to architects and engineers may also play a role. These include the American Institute of Architects (AIA) and the Institute of Electrical and Electronics Engineers (IEEE).

- Builders, electricians and other trade allies will be anxious to help facilitate the expansion of the integration between the power delivery system as it exists today and the fully functional power delivery system , as it may exist tomorrow. The explosion in these areas infers the demand for practitioners with enhanced need for skills not now replete in the existing electricity enterprise. Organizations and associations which support these groups should be particularly supportive, including the International Brotherhood of Electrical Workers (IBEW) and the National Electrical Contractors Association (NECA).

• Academicians are typically sources of the most cutting-edge thinking in our society. We need to engage the research minds at universities to contribute to the understanding of consumer responsiveness to electricity issues, including price.

DETERMINE STATE OF INDUSTRY AND
TECHNOLOGY INCLUDING DRIVERS AND BARRIERS

DR-Ready specification must be informed by the state of existing product features through input from vendors and the manufacturing industry. A collaborative effort is required with industry to identify an achievable specification that can be met with a broad range of products in a specific area. Ideally the DR-Ready specification would be neutral to a specific technology and ahead of the curve, so as to lead the industry in the direction of public interest.

Key drivers for DR include peak resource constraints, improved economics, reliability concerns, environmental concerns and enhanced innovation. Among the barriers to be explored, the following may exist: technological barriers, aggregation, automation, system operator confidence, economic justification, wholesale market structures and retail rates, and customer convenience. Ideally product specifications would help overcome existing challenges towards achieving the benefits associated with the key drivers.

DR functionality does not need to be a stand-alone feature. If it is, purchasers will likely not pay much attention to this for most products. However, network connectivity—especially if it is done with universal interoperability—can provide many features that are of interest to people. It may be key to package DR with other features so that people get it frequently but without their asking for it. As an example, organizations such as Digital Living Network Alliance (DLNA) reference collections of standards that, if all are implemented, should result in products that are interoperable and provide good user experiences. EPRI (and DOE, EPA and NIST) could seek to wire DR into efforts like DLNA to provide entrée to many products, and critically, to other requirements that enable interoperability. Such other requirements are a needed part of enabling networked DR an so cannot be ignored in any case.

In terms of selling DR to people, having features that are much more attractive than DR itself may be key. An example of this is the related areas of distributed generation, off-grid generation and emergency situations. As more people have some ability to generate power locally, from UPS systems, hybrid cars or solar panels, they may want or need to have the ability to manage local consumption as if it was a grid, with a local "price" to ensure that demand does not outstrip supply. DR for on-grid operation could be packaged as part of a feature set that enables off-grid operation, particularly for emergency situations like earthquakes or weather-related catastrophes.

DEFINE PRODUCT ATTRIBUTES THAT WARRANT "DR-READY" DESIGNATION

For example, these attributes could include requirements that the end-use device or appliance or information technology and consumer electronics must include a communications interface, automated controls and plug-and-play installation capability.

A critical issue for network architecture is the question of "locus of authority." That is, which device has some say, and the final say, on what a particular device does regarding DR or price responsiveness. Is it the device itself? Is it the grid? The occupant? A home control device? A home gateway (to the internet) device? The electric meter? There isn't one universal answer, but the topic is clearly one that needs research in the process of designing the overall network architecture.

An *integrated two-way communications* interface would enable the ability of the system operator to control the end-use device and to be alerted to its status. This would include allowing acknowledgement signals to be relayed back to the utility or aggregator. Control and acknowledgement permit continuous monitoring and assessment of resource visibility and availability.

Automated controls permit configurable preferences for default demand response actions to be automated resulting in rapid actuation of response. Dimmable and curtailable settings allow either full interruption (e.g., for pool pump, air conditioner, water heater, etc.) or partial reduction (e.g., dimmable lighting, consumer electronics with multiple power modes, and other curtailable end uses).

Plug and play installation capability embodies a concept analogous to the capability which Microsoft® operating systems now use. Today when you plug in a new printer or other hardware, the system automatically organizes itself to enable the device. In order for this capability to exist, communications standards must be supported (e.g., Zigbee, Homeplug, etc.). The capability must be DR program agnostic and allow configurable participation parameters.

Standardization of user interface elements related to DR and price responsiveness is critical. A good reference for how this has been done in related areas is the Institute of Electrical and Electronics Engineers (IEEE) Standard IEEE 1621. For more on that, see http://eetd.LBL.gov/Controls.

BUILD COALITION WITH STRATEGIC PARTNERS

A multi-stakeholder process must be launched in order to establish minimum requirements definition. This definition may include outlining a "common information model" (CIM) for seamless connectivity across applications. The CIM could be software which allows input from any source to be matched with any communications protocol. CIM concepts have been effectively applied in system operations and wholesale market operations.

The first step in this assessment could be a joint EPA/DOE/EPRI workshop with manufacturers. This would allow the core coalition to understand vendor interests, concerns and barriers, as well as their interest in selective demonstration opportunities and joint development and pilot demonstrations. An early outcome of this workshop would allow the identification of gaps and prioritization of RD&D. The core coalition would subsequently coordinate demonstrations with manufacturers. They would connect with manufacturers in the specification phase. Through this process a familiarity would be developed with the state of product capabilities, technical and business challenges as well as research needs and gaps

DEVELOP ROADMAP OF TARGET PRODUCTS

A roadmap would be developed of targeted products in both time and development capability. This would start with one or two key appliances, based on energy use, the state of technology, and ease of implementation. High-usage devices/appliances (large MW/MWh savings potential) would have priority. In order to determine this priority analytics need to be developed to quantify benefits of DR-ready by end use.

It is likely that DR-ready would first be applied to either thermostats or other controllers tied to central air conditioners and heat pumps. Later phases would potentially address the appliances and devices listed in Table 12-7.

DEVELOP STANDARDS FOR EXCHANGING INFORMATION WITH SMART APPLIANCES

This last step is complex and contentious. It reflects the realization that communications systems will evolve so as to become overlaid with electric distribution systems. As this occurs, standards will need to be developed which define a standard data packet for two-way communications and data exchange. This packet design must be media-agnostic and permit wireless, fiber optic, twisted pair, broadband over power line or other physical media to operate independently or in conjunction with one another.

This two-way communications must be enabled so as to allow the Smart- or Intelligent Grid to be connected in a media-agnostic way with the consumer's building through an advanced metering infrastructure or other consumer portal. This connection must be enabled so as to allow subsequent, seamless integration with intelligent communicating end-use devices and/or appliances.

Table 12-7. Candidate Product Areas for DR Ready Designation

Room air conditioner	Furnaces and boilers
Standby generation	Electric water heaters
Refrigerators, freezers and refrigerator/freezers	Residential clothes dryers
Thermal storage systems	Battery chargers (including those for electrically powered vehicles)

In addition, the DR-ready approach must standardize messages that will be sent to smart devices. This will include a communications address like an IP address and a defined network "envelope."

This network envelope would contain an application layer with a defined data model including device identify and information such as price. Price information would further be divided into functional requirements like schedule, units, dates, times and hourly prices.

References

"Benefits of Demand Response in Electricity Markets and Recommendations for Achieving Them," a report to the U.S. Congress by U.S. Department of Energy, February 2006.

"Assessment of Demand Response and Advanced Metering," Staff Report, Federal Energy Regulatory Commission (FERC), December 2008.

"Assessment of Achievable Potential from Energy Efficiency and Demand Response Programs in the U.S.: (2010-2030)," EPRI, Palo Alto, CA: 2009. 1016987.

"Advancing the Efficiency of Electric Utilization: Prices to Devices[SM]," Background Paper, EPRI Summer Seminar, 2006.

"Appliance Interface for Grid Responses," C. Eustis, G.R. Horst and D.J. Hammerstrom, *Proc. of the Grid Interop Forum*, November 7-9, 2007.

"Networks in Buildings: Which Path Forward?", B. Nordman, American Council for an Energy Efficient Economy (ACEEE), Scaling Up: Building Tomorrow's Solutions, 2008 ACEEE Summer Study on Energy Efficiency in Buildings, August 17-22, 2008.

"Development and Demonstration of Energy Management Control Strategies for Automated Real-Time Pricing," EPRICSG, Palo Alto, CA, New York State Energy Research and Development Authority, Albany, NY, Consolidated Edison Company of New York, Buchanan, NY, and Empire State Electric Energy Research Corporation, New York, NY: 1998. TR 111365.

"Automated CO_2 and VOC-Based Control of Ventilation Systems Under Real-Time Pricing," EPRICSG, Palo Alto, CA, New York State Energy Research and Development Authority, Albany, NY, Consolidated Edison of New York, Buchanan, NY, and Empire State Electric Energy Research Corporation, New York, NY: 1998, TR-109117.

Index

Printed and bound by CPI Group (UK) Ltd, Croydon, CR0 4YY

23/10/2024

01777696-0009